*Studies in Consciousness / Russell Targ Editions*

Some of the twentieth century's best texts on the scientific study of consciousness are out of print, hard to find, and unknown to most readers; yet they are still of great importance. Their insights into human consciousness and its dynamics are still valuable and vital. Hampton Roads Publishing Company—in partnership with physicist and consciousness research pioneer Russell Targ—is proud to bring some of these texts back into print, introducing classics in the fields of science and consciousness studies to a new generation of readers. Upcoming titles in the *Studies in Consciousness* series will cover such perennially exciting topics as telepathy, astral projection, the after-death survival of consciousness, psychic abilities, long-distance hypnosis, and more.

BOOKS IN THE STUDIES IN CONSCIOUSNESS SERIES

*An Experiment with Time* by J. W. Dunne

*Mental Radio* by Upton Sinclair

*Human Personality and Its Survival of Bodily Death* by F. W. H. Myers

*Mind to Mind* by René Warcollier

*Experiments in Mental Suggestion* by L. L. Vasiliev

*Mind at Large* edited by Charles T. Tart, Harold E. Puthoff, and Russell Targ

*Dream Telepathy* by Montague Ullman, M.D., and Stanley Krippner, Ph.D., with Alan Vaughan

*Distant Mental Influence* by William Braud, Ph.D.

*Thoughts Through Space* by Sir Hubert Wilkins and Harold M. Sherman

*The Future and Beyond* by H. F. Saltmarsh

RUSSELL TARG EDITIONS

*UFOs and the National Security State* by Richard M. Dolan

*The Heart of the Internet* by Jacques Vallee, Ph.D.

STUDIES IN CONSCIOUSNESS
*Russell Targ Editions*

# Thoughts Through Space

## A Remarkable Adventure in the Realm of Mind

**Sir Hubert Wilkins
Harold M. Sherman**

Originally published in 1951 by C & R Anthony, Inc.
Original copyright © 1951 by Sir Hubert Wilkins and
Harold M. Sherman.

Foreword by Ingo Swann copyright © 2004
Cover photo collage by Mayapriya Long
Cover design by Bookwrights Design

All rights reserved, including the right to reproduce this
work in any form whatsoever, without permission
in writing from the publisher, except for brief passages
in connection with a review.

Hampton Roads Publishing Company, Inc.
1125 Stoney Ridge Road
Charlottesville, VA 22902

434-296-2772
fax: 434-296-5096
e-mail: hrpc@hrpub.com
www.hrpub.com

If you are unable to order this book from your local
bookseller, you may order directly from the publisher.
Call 1-800-766-8009, toll-free.

Library of Congress Cataloging-in-Publication Data

Wilkins, George H. (George Hubert), Sir, 1888-1958.
  Thoughts through space : a remarkable adventure in the realm of mind /
Sir Hubert Wilkins and Harold M. Sherman.
        p. cm. -- (Studies in consciousness)
Originally published: Hollywood : House-Warven, 1951. With new introd.
  ISBN 1-57174-314-6 (6x9 TP : alk. paper)
  1. Telepathy. 2. Arctic regions. I. Sherman, Harold Morrow, 1898-
II. Title. III. Series.
BF1171.W49 2004
133.8'2--dc22
                                                    2003025269

ISBN 1-57174-314-6
10 9 8 7 6 5 4 3 2 1
Printed on acid-free paper in Canada

*In memory of
Sir Hubert Wilkins
and Harold Sherman,
intrepid explorers both.*

# Contents

FOREWORD
by Ingo Swann ................................................. ix

THE EVIDENCE FOR TELEPATHY
by Sir Hubert Wilkins ........................................... xv

LOOKING BACKWARD—AND FORWARD
by Harold Sherman ............................................. xix

## PART ONE: MY STORY AS SENDER—WILKINS

1. I ARRANGE TO EXPLORE THE MIND ........................... 3
2. ENROUTE TO THE FAR NORTH ................................ 13
3. SOME SIGNIFICANT TELEPATHIC "HITS" ...................... 26
4. A FIRE IN POINT BARROW ................................... 41
5. HIGH EMOTIONAL MOMENTS .................................. 66
6. SHERMAN'S TELEPATHIC RECORD ............................. 88

## PART TWO: MY STORY AS RECEIVER—SHERMAN

1. WHAT HAPPENED TO ME ...................................... 99
2. HOW THOUGHTS ARE RECEIVED ............................... 114

3. THE PART EMOTIONS PLAY .................................127

4. UNKNOWN POWERS OF MIND ............................133

5. KEEPING ONE'S MENTAL AND PHYSICAL BALANCE .........139

6. PRACTICAL USE OF TELEPATHY ............................144

**PART THREE: AUTHENTICATED DOCUMENTARY RECORD OF THE WILKINS-SHERMAN EXPERIMENTS IN LONG-DISTANCE TELEPATHY** ...............................151

# Foreword

This remarkable book was published in 1942 during the escalating urgencies of World War II.

At first it attracted a fair amount of media attention, largely because of the worldwide fame of Sir Hubert Wilkins. But during those times of wretched war and turmoil, fresh evidence for long-distance mind-to-mind interaction could not top out on must-read lists. And so interest in the book quickly dwindled, even in parapsychological circles.

Because of those unfortunate circumstances, the dramatic experiments of Wilkins-Sherman never really received their appropriate place among the annals of significant displays of human powers of mind.

However, some sixty years later and through the good works of Russell Targ, this book is now drawn up out of historical dustbins to finally find a rightful place in this important series of significant studies in consciousness.

Many developments have taken place during those six decades, and some of them help shed new luminosity upon the meaning and value of the Wilkins-Sherman experiments.

And so, although the book itself is very carefully constructed and self-explanatory throughout, and very easy to read, an additional foreword is desirable.

The first thing to be done in this foreword is to refresh the personalities of the two principle men involved.

Sir Hubert Wilkins (1888–1958) is largely forgotten now. But back in the 1920s–1940s anyone with access to newspapers and radio certainly understood that he was a larger-than-life figure, a superhero explorer and aviator, so much so that the media kept track of his every activity.

Born in Australia, Wilkins enlisted with the Australian Flying Corps in World War I, and afterward served as photographer and naturalist in several polar expeditions. In 1926 and 1927 he commanded two expeditions to the Arctic.

A pioneer in the methods of air exploration, in 1928 he was the first to fly from North America to the European polar regions, and in a long flight of more than twenty hours, he succeeded in proving that there was not a continent under the arctic ice. He was knighted for that achievement.

In 1928–29, he commanded an expedition to the Antarctic. In 1931, in a submarine named *Nautilus,* he headed an under-ice expedition to reach the North Pole, and his work was to be very valuable for future Arctic exploration by submarine. From 1933 to 1938, he was manager for Lincoln Ellsworth's expeditions across the Antarctic continent.

During 1938, responding to official Russian requests, Wilkins searched for a group of Russian aviators lost in the Arctic. It was during this search that the attempts were made by him and Sherman to communicate by mental telepathy. During and after World War II, Wilkins served as a geographer for the British army.

Thus, in the years prior to World War II, Sir Hubert Wilkins was among the most preeminent and daring explorers of the world as it was back then. His exploits and reputation were secure in those contexts, and so there was little reason for him to mix into, of all things, telepathic experiments across vast distances.

Harold Sherman (1898–1987) was born in Traverse City, Michigan, educated at the University of Michigan, and lived in New York before he and his lovely wife, Martha, settled into humble premises on 200 acres of near-wilderness in Arkansas. He authored sixty books and a number of plays. His many memberships included the Authors League of America, and Dramatists Guild, and he consulted on several Hollywood screenplays.

He was also well known as a reporter and radio commentator, and was ultimately the author of several popular "how to" books to guide people in the development of their ESP.

His deep interests in extrasensory perception ultimately eclipsed his other endeavors. In the 1920s and 1930s, he conducted experiments in clairvoyance, telepathy, psychokinesis, precognition, mediumship, and survival after death of the physical body.

But Sherman went one important step beyond mere experimenting. With psychologist Leslie LeCron and other scientists affiliated with the University of California in Los Angeles, he began investigating "methods of operation of the ESP faculty"—i.e., how mental processes of the faculty worked.

How Wilkins and Sherman got together is narrated in the text ahead. And thereafter follows the fascinating descriptions and results of six months of their otherwise unlikely 1938 collaboration in long-distance telepathy between New York and the Arctic.

In early 1970, I discovered a used copy of this book in the basement of Weiser's book store in New York, and I was thereby amazed and staggered not

# Foreword

only by the telepathic successes described in it, but by the many implications for the real existence of extraordinary forms of consciousness.

I really wanted to know Harold Sherman, but when I asked about him in parapsychology circles no one knew very much—except to suppose he had died "years ago."

It was in 1971 that I had begun acting as an experimental subject for Dr. Karlis Osis and Janet Lee Mitchell at the American Society for Psychical Research where I had given many rave reviews of the Arctic experiments, and expressed regrets that Sherman had died.

One day in November of 1972, when I reported for research duties, Janet, her eyes gleaming, said: "Well, you are about to get one of the wishes of your life. Harold Sherman called here. He is in town seeing his publisher. He's staying at the Algonquin Hotel, and, if you don't object, he very much wants to meet you."

Janet and I went along to the Algonquin, the famous watering hole for East Coast intellectuals of all kinds.

As we soon discovered, Harold Sherman, then in his 74th year, turned out to be a refined gentleman in all ways, patrician in bearing and dress, and courteous in all situations.

It did not take long to also see he was very "world wise," but with no trace of being "world weary." And, as time would reveal, he had hardly a shred of ego, and was genuinely interested in others. He was also a virtual library regarding all matters parapsychological both up front and behind the scenes. And he was certainly not dead, as rumor had it, but very much alive, kicking, and dynamic in more ways than one.

Two of the enormous values of this remarkable book can be missed if they are not pointed out.

First, it documents a "telepathic" experiment that consistently went on across great distances for no less than *half a year*, including preparation and post-analysis time. And in that sense it was a tour-de-force feat of exploration that had never been conceptualized or undertaken before.

Second, although the term "telepathy" is used to describe what went on, a careful reading of the text shows that term to be insufficient, principally because it limits recognition of factors that are not normally included in its established meanings.

As the reader will soon see, when Wilkins finally arrived in the Arctic, he was assailed by urgent and hazardous difficulties of all kinds coming from extreme weather conditions and equipment failures. There is no evidence in the book establishing that Wilkins ever actively took time to "send" telepathic information to Sherman who passively "received" what was sent. What he did do was to record events and details in his log, this being a usual habit with an explorer. When Wilkins returned from the Arctic, his dated log was compared with the dated impressions of Sherman.

I asked Harold about this aspect during the early months of our very good friendship. "Oh, my goodness no," he replied. "We knew from the start that Sir Hubert probably would be too stressed and too fatigued to send anything on the appointed days when I would try to be in contact with him and where he was."

He added: "Sir Hubert had exceptional intuition himself, but was unsure about his own telepathic faculties."

So I asked Harold why the concepts of telepathy, mind-to-mind, and thoughts through space seemed so pronounced in the book. His answer was so astonishing and revelatory that it opened up an entire new vista for me.

"Well," he began, "this has a lot to do with the title of the book. Sir Hubert was interested in rapport with things and surroundings from which he drew his intuitions. We originally wanted a title along the lines of rapport across space and time. Rapport had fallen out of fashion in parapsychology. So in the publishing process it was felt that the telepathic idea of thoughts through space was more up to date, more recognizable to the public, and would serve better."

Briefly defined, *rapport* is "relationship marked by harmony, conformity, accord, or affinity."

And if anything can be deduced from this book, it is that it was this kind of relationship the two authors shared with each other.

In any event, what *is* unavoidably evident in the book is that Sherman *did* "tune in" to the stressed out Wilkins via some kind of rapport that at least contained affinity, harmony, and accord.

From this rapport state, Wilkins "downloaded" information elements to Sherman that could be categorized as telepathy or thought transference through space, as traveling clairvoyance, and as empathic and emotional joining. The mix of all these factors enabled Sherman not only to focus on Wilkins himself, but on environmental factors and things around him.

It might also be noticed that the rapport state also occasionally enabled Sherman to pick up on some intentions of Wilkins, but which were never carried out, and to pick up on meanings for which there was no direct physical or situational evidence.

All of this implies the existence of a state or states of "consciousness" that have not really been identified before, at least with respect to perceptions across thousands of miles.

During the years of my friendship with him, Harold Sherman completely agreed with the considerations briefly outlined above.

He often initiated discussion of them on his own steam, and recounted that he and Wilkins had discussed them both before and after their long-distance experiments.

He pointed out that the traditional categories of phenomena set up to describe various modes of paranormal perception were inadequate in the face of states of consciousness that had not yet been identified.

It was because of this that Sherman took to the concept of remote viewing like a duck takes to water. He said that Wilkins would have done likewise—the reason being that both he and Wilkins were interested in new knowledge rather than in forcing phenomena into the self-limiting definitions of old terminology.

The concept of "remote viewing" came into existence only in late 1971, and so it was not available in 1938. However, that concept proved to have one advantage.

If one desires to get information about a distant place, then the *getting* of the information is more important than are the nomenclature categories used to get it.

Indeed, in parapsychological experiments it often proved difficult to tell if the information was obtained via the nomenclature categories of telepathy, clairvoyance, or precognition, etc. The contexts of remote viewing had the advantage of focusing on information to be obtained without specific nomenclature reference as to how it was obtained.

One kind of experiment that came out of the remote viewing context was the "out-bound" experiment. An individual went to a distant location. The subject didn't have a clue as to where or what the site was, but was supposed to view it anyway.

This type of experiment was intensely researched and developed by Hal Puthoff, Russell Targ, and others at Stanford Research Institute (SRI) in the early 1970s.

One very intriguing result of this large, government-sponsored, SRI effort involving approximately 150 volunteers showed that even disbelieving subjects did get bits and pieces of information about the distant location the out-bound person went to. This could be interpreted to imply that everyone had ESP faculties, whether they realized it or not.

Harold would surely want me to say, in this foreword, that this idea was fundamental to his heart and life, and that it accounts for his many "how-to" books such as *How to Make ESP Work for You* (1964).

During the early decades of the twentieth century there had been some experimental efforts of the out-bound type. But in 1938, with Wilkins out-bound into the remote, unforgiving vastness of the Arctic and without a cell phone, and Sherman as remote viewer in New York some thousands of miles distant, a new benchmark for potential "remote-viewing" *consciousness* and efficiency was opened up and demonstrated in depth and in amazing detail.

**Ingo Swann**
**New York, 2002**

# The Evidence for Telepathy

## By Sir Hubert Wilkins

THE ACCEPTANCE of evidence as factual and the scientific establishment of the fact are by no means simultaneous. Just recently I was confronted with a rather amusing example. In 1913, already able to fly an airplane, I went to the Arctic and there walked many thousands of miles over the frozen sea, keeping a pilot's eye open for places suitable for the landing of an airplane. As a result I concluded that such would be possible within a radius of 25 miles from almost any point on the Arctic sea ice, and so reported. I was asked, "But did you land an airplane upon the ice?" We had no airplane with us on that expedition and, naturally, the answer was in the negative. Then, said my scientific friends, your statement about landing planes on the ice is merely an assumption.

Later, during the years 1926–28 and 1937–38 the pilots associated with my expeditions landed planes many, many times upon the Arctic sea ice and these landings were widely reported in the newspapers of the world. Following these flights and when it became likely that military planes might be required to land on the ice of the Arctic sea, the evidence I had accumulated was presented as proof of the safety of such. The scientists then asked "Were your landings directed, observed and studied by scientific personnel?" Since usually there were only two of us in the plane and often there were no observers, I could not say that all the landings and takeoffs we safely made were under scientific observation or control. So the verdict of the savants was *"You* say that landings and takeoffs were safely made, but that is only your opinion." I am, however, alive to prove it. Then, this year, 1951, two military planes under official orders made half a dozen landings upon the Arctic sea ice and the verdict was proclaimed "It has now been established as a fact that an airplane can land and takeoff from the Arctic sea ice."

There has been a far greater span of time since the first demonstration of

Extra-Sensory-Perception, or thought transference and the recordings of Harold Sherman than the interval between 1913 and 1951. Adam may have used thought transference to communicate with Eve and it seems likely that the wisdom of the Serpent was not entirely conveyed to Eve by means of human colloquial dialogue. But still, communication by means of thought power alone between two people who were, comparatively, strangers, over a distance of 3400 miles and established as authentic beyond doubt has not, apparently, had much effect on the acceptance of telepathy as a scientific fact.

Now, almost ten years after the original publication of "Thoughts Through Space" and after re-reading the manuscript, it seems to me that while the tale is told in straightforward simplicity, there lies within it a rousing, staggering, indication of potential powers of the mind, and a signpost toward not necessarily the use of mind for the purpose of thought transference, but toward the potential of mind development and the wider and better uses to which it may be put.

Recognition will surely come, in time, and how little we of today comprehend time and the essence. In the text may be found an example of completed thought and the time element in regard to its effect. Effect, apparently may be delayed, or in some few clairvoyant cases it seems to outrun conscious recognition by its maker. In the book may be found evidence that "concentration of thought" and "deep thought" in the accepted sense is not necessary for thought "power." Neither need the thought be of momentous importance to be impressive. So far as response is concerned, such trivialities as selecting fish for dinner were as surely registered as was the noblest of thoughts in regard to the progress of civilization.

So far as the energizing factor is concerned, it requires no tense-browed concentro-determination that the thought will have effect; rather it seemed that simple faith was the most effective dynamo in the matter of transmission and response. But do we really know what we are thinking? How much so-called subconscious effort takes place without reflection on the surface of recognition? In the text may be found several instances of subconscious emanations as clearly pictured by Sherman as those emotional incidents which seemed to absorb my physical attention to the exclusion of all else, although an analysis will indicate that it was the emotional incidents which, in general, were selected by Sherman for his records.

At times Sherman seemed to sense a feeling of intensity, especially when there was simultaneous co-operation. Then there is the phenomena of clarity of vision. It seems odd that Sherman viewing my features on his screen of recognition would detect upon my cheek a scar, no bigger than a pea, which he had failed to notice during the several times we had sat vis-a-vis in broad daylight, or under strong electrical illumination. Since the printing of the first edition of this book circumstances have prevented a sustained effort on our part to further the accuracy and consistency of the phenomena here recorded, but I for one am continually amazed at the apathy (my own included) exhibited in respect to the potentials of mind development and control.

Man has found the tools with which to handle the intricacies of nuclear physics; perhaps the basic principles of the mind are more evasive and intricately involved because they are not entirely physical, and the mind of man has not yet come to accept the spiritual as a field offering desired reward for research capital invested. But doubtless the time will come when a verdict will announce "It has now been proven that thought communication, without mechanical aid, is available to those whose minds are trained and whose will is directed toward the betterment of mankind."

# Looking Backward—And Forward

By Harold Sherman

LOOKING BACK NOW, along the measuring stick of Time, to my pioneering experiments in what was popularly called "long distance telepathy," I am amazed at the growing significance of the work in which Sir Hubert Wilkins and I engaged. We had entered this unique adventure with the enthusiasm of two men who shared a serious interest in the possibility of communication, mind to mind, but with no substantial expectation of achieving any consistent success. The best that could be hoped for, we thought, might be a number of "coincidences," wherein impressions recorded by me, as the receiver, would seem to parallel actual experiences undergone by Wilkins. Even such an achievement, in our opinion, would have been worth the protracted effort of experimentation which, as it turned out, extended over a continuous five month period, three nights a week.

For me to have had the "inner vision" of a fire, for instance, at the very moment when Wilkins, more than three thousand miles away, was actually seeing an Eskimo's house ablaze, would have been evidence enough that something beyond "mere coincidence" was at work. The laws of chance against my exactly determining at any given moment what was happening to Wilkins or what event he might be witnessing were astronomical. Yet, to the increasing surprise of both Wilkins and myself, as our experiments progressed, it was found that my recorded impressions were maintaining a high percentage of accuracy.

Not only that but the nature of the impressions received interested us greatly. Many were extremely personal and not directly related to the search expedition itself. Some impressions, equally correct, had not come from the mind of Wilkins, and had to be checked by him to prove their veracity. There was the time that I recorded an episode in the life of a member of Wilkins' crew. Wilkins, himself, might never have known of it, had it not been for my "pick-up" of the event and his subsequent investigation.

Where did such an impression come from? Had my mind received it from the mind of the crew member who, in trying to conceal this misadventure, was nevertheless unconsciously "broadcasting" it?

Did this indicate that (as I now firmly believe) all humans are broadcasting and receiving instruments, and that a sensitively attuned individual may, at times, tune in on the thought forms of others, at any distance?

Certainly, when I sat quietly in my study in New York, and directed my mind to "determine for me what is happening to Wilkins at this moment or has happened to him THIS day," I was getting a report back in the form of flashing mental pictures and strong feelings which, for the most part, were found to have a corresponding basis in fact.

How my mind, or any mind could function in this manner, was a mystery to me, and remains a mystery yet today. But I was able to demonstrate, as have many others before and since, that if I made my mind properly receptive, I could "become aware" of what was happening at a distance or of what others were thinking.

I discovered that the regular practice of this "perceptive faculty" of mind definitely contributed to its greater effectiveness. Like a muscle of the body, which, persistently exercised, develops increasing strength, the sensitivity of my mind seemed to improve as I made concentrated demands upon it.

My two greatest obstacles to accurate receiving were the difficulties I encountered in shutting off my imagination and in controlling my natural doubts and apprehensions. This had to be done each time I kept my appointment with Wilkins and I can charge any failures in reception quite largely to the times that I allowed my imagination or my doubts and fears to "color" what was "coming through."

Present and future experimenters will have these same obstacles with which to contend. The imaginative faculty, unless rigidly curtailed, likes to embroider each impression. So subtle is its operation, you must develop a fine and almost instantaneous sense of discrimination, to be able to detect when it is seeking to enlarge upon the impressions received.

Doubts and fears, rising out of one's emotional nature, can have just as distorting an effect upon the receiving of thought forms—and these, too, must be kept from intruding their inhibiting influence.

Performance of telepathy is not, in itself, too difficult. Any earnest experimenter can demonstrate to his own satisfaction, after a little conscientious practice, that thoughts can be sent and received. But he must, at the outset, accept on faith that it is possible to communicate telepathically. If he does not, his own skepticism will prevent this faculty of mind from functioning, particularly if he is to act as the receiver.

One may receive thought impressions from the mind of a skeptic or a nonbeliever in the existence of telepathy, but it is extremely difficult for any doubting person to successfully receive. There is apparently a scientific reason for this.

## Looking Backward—And Forward

Your Subconscious mind operates under direction of your Conscious mind. It takes its orders in the form of mental pictures or feelings from the Conscious mind. Whatever you doubt or fear or desire is registered in consciousness with an intensity equal to the intensity of your own emotions at the time—and this constitutes your mental attitude on this subject or experience, until and if it may be changed by your future reactions.

For this reason, if you tell yourself, with conviction, that there is no such thing as telepathy, before you undertake to act as receiver, you are simply saying to this inner faculty, "Don't function for me"—and it doesn't.

Years ago, when I was first experimenting with telepathy, I had to assume that it could be done and that I could do it. I had to charge my consciousness, so to speak, with the faith that it could and would bring to me accurate impressions from the minds of others. It was necessary for me to work up, in my mind, an eager, vital interest in the undertaking and a keen sense of anticipation. I would "wonder" what impressions my mind was going to bring to me and mentally await, with a high degree of inner expectancy, the delivery to me of these impressions. And when they flashed across my consciousness in the form of fleeting pictures or feelings, it was my task to capture them by writing down what I thought these pictures or feelings meant, in my own words. Sometimes the pictures or feelings were so faint or fragmentary that I was unable to catch much, if any, of them. At other times they etched themselves so strongly in consciousness, for the moment, I was enabled to record their nature quite fully. And, at these times, the vivid impressions were usually accompanied by a simultaneous, indescribable feeling in my solar plexus which said to me: "This is it! This is true! You've got it!"

Savants have called the solar plexus, this great nerve center of the body, man's "second brain." We don't begin to understand, as yet, its mysteries and complexities, any more than we can comprehend the mysteries and complexities of our own minds—but there is undoubtedly a vital connection between the two.

Humans testify that when sudden fear or apprehension has struck them, they have felt a "sinking feeling in their stomachs," "a pain or tightness in the center of their bodies," as if they have been "hit a blow there." It is as though one's thoughts and emotional reactions are grounded in the solar plexus. As one sensitive individual has put it, "I seem to perceive with my mind and feel with my stomach."

Whatever the connection between mind and solar plexus, this much is certain, as Wilkins himself has observed: The intensity of a sender's emotional reaction to what is happening to him, or has happened, determines the degree of intensity of the "thought waves" discharged.

As a receiver of Wilkins' thoughts, I noted that I received most easily and vividly, impressions concerning experiences which had moved Wilkins deeply, emotionally.

A fire, a severe toothache, a narrow escape from death in his plane, the wearing of a "loaned evening dress suit"—all these types of emotionalized happenings registered in my consciousness so distinctly that I was able to describe them with unmistakable accuracy.

Some day, when research has determined the fundamental laws having to do with communication between minds, I am sure that one's feelings or emotions will be established as the generators of power—the transmitters of "mind energy"—the force behind the impulses which fly from one mind to another.

In the files of psychic research societies are thousands of cases describing impressions people have received of tragic happenings to friends and loved ones. These impressions may have flashed into consciousness while the individual was awake or sleeping. Many have dreamed that a son, daughter, husband, wife or sweetheart was in a serious accident or had died in some manner, only to learn later that this was indeed true.

When you consider that a human, seriously hurt or dying, is ordinarily in a high emotional state and would naturally be thinking strongly of loved ones—it is not difficult to understand how this powerful feeling might project thoughts with great intensity through space. Under such conditions, unless the object of these thoughts was mentally preoccupied at the moment, there would be every chance of these outer impressions registering in consciousness. They might not command the attention of the individual vitally enough to give him an exact sensing of what has happened to a loved one—but would have sufficient impact, nevertheless, to make him feel uneasy and, possibly cause him to send a wire or make a long distance phone call of inquiry.

"I have a feeling that things aren't right or aren't going so well with So-and-So," many have said. Later, they discover that their feelings were correct. They had somehow managed to "tune in" on the emotional disturbance in the mind of the friend or relative. Either that, or the "emotionally energized thoughts" of this person had gained access to their own minds.

We still stand in the presence of phenomena without being able to explain it. We know, under certain conditions, thoughts can be sent and received. No genuine "sensitive" can guarantee a hundred per cent accuracy in any experiment. There are times when reception approaches this point but these are the rare exceptions rather than the rule. If one tries to "force" the receiving of impressions, his score of "hits" is usually close to zero. One must be physically and mentally relaxed, quiescent, to best receive impressions. As nearly as the receptive state may be described—it is an "inner alertness" and an "outward passivity."

Should you experiment with what is now called Extra Sensory Perception—you will find that more than telepathy is involved in your sensing of thoughts and conditions. At times your mind will pick up impressions of past and future events, both seeming to be co-existent insofar as awareness is concerned. I sensed and accurately recorded, the only two accidents which were

# Looking Backward—And Forward    xxiii

to befall Wilkins' plane in the Far North, at least ten days in advance of the actual happenings. In these instances, I certainly was not receiving this knowledge from the mind of Wilkins. And yet, some intelligence within my own consciousness, most probably, had "reached ahead in time" and pictured for me events which had not yet occurred.

There is a growing abundance of evidence now that precognition is a fact. Man's mind can go backward and forward in Time, with equal facility, when so moved to function.

This rightly raises the dual question: Just what *are* Time and Space? Perhaps our limited concepts of Time and Space inhibit us from properly utilizing our mindal powers with respect to these dimensions. Occasionally, some form of receptivity or some compelling impulse, from within or without, causes us to transcend or step outside and beyond our ordinary mental limitations, and grasp, in a flashing instant, some comprehension of a happening, past, present or future, of which we would not normally be aware.

We may eventually learn that all thoughts have a rate and character of vibration and continue to exist around and about us, after creation and emanation from the mind. If this is so, then, theoretically, a "sensitive" individual might tune in on thoughts, particularly those which have been highly emotionalized, days, weeks, or even years after their discharge.

Certainly, all thoughts you have had are now coexistent in the memory stream of your own consciousness, otherwise you could not recall many of them at will. This no doubt explains why some true "mind readers" can accurately recite the outstanding happenings in an individual's life. I, myself, have suddenly "felt that I was another person," momentarily, and in that moment have been able to describe certain feelings and experiences of which I was then made aware, just as though I was that individual, recalling them. Nor had I ever seen these persons before, in many instances, it being impossible for me to have gained any knowledge of their past through any physical means.

How ironic it is that Man, with all his material and scientific accomplishments, still remains in such abysmal ignorance of his own self—his own mind and body!

Were some of the billions now expended for war, diverted to the study of Man, himself, it is my conviction, shared by Sir Hubert Wilkins, that the causes of war itself, could be removed.

Man's mind and how he uses it, creates his world and also destroys it. Man possesses undreamed of powers within his own consciousness which, properly developed and applied, could enable Man to solve all his worldly problems and arrive at a true understanding of himself and his fellow man.

Until and unless Man learns to make constructive use of his "thinking machine," he will continue to bring more and more devastating calamities upon himself.

To those who ask, "what practical value is to be served by a study of the mind?" we answer: "How Man thinks and feels determines what he is. Thought is the most vital force in the universe. Right thinking can change the entire course of an individual's life and, if enough humans think right, the course of a nation and the world, itself, can be changed. This being true, Man must eventually turn to himself for the key to his own liberation—otherwise his wrong thinking will enslave himself and the world."

Our experiments convinced Wilkins and myself that humans are still in the kindergarten of their own mental processes—that few of us yet begin to sense our possibilities for development and attainment—that Man has a much higher destiny than he has realized or imagined—and that the Great Unknown can be more and more knowable if Man is willing to pioneer in the great adventure of seeking to gain a finer and finer understanding of his own self and his true relation to this wonderful Universe.

Part One

# My Story As Sender

### By Sir Hubert Wilkins

# 1

# I Arrange to Explore the Mind

A PALE OLD MAN sat in a small room in a remote village in Poland. Before him was a chart of the North Polar Regions. He was, apparently, staring into space. Suddenly he was moved to action.

"Ice, ice, ice," he shouted, "ice everywhere. What enormous spaces! My Lord, what greatness! What a wonderful scene! Immense fields of ice in white and bluish shades. There they are," and he marked with his pencil a small circle on the map. "The ice is moving rapidly, something is forcing it about, shoving it away. But men are there—all of them—alive. Their radio has failed and they are unable to let us know. Some kind of animal is there with them. Did they have a dog? Two men are injured—one seriously. Their airplane gives the impression that the landing gear is out of order. Levanevsky is holding up the morale of the whole crew, he is unusually active. One of them is trying to repair the radio. Their greatest danger is the rapid movement of the ice. They are now about Longitude 170 E., but they landed at about Longitude 165 W. Those who are well refuse to leave the plane and their sick comrades. The radio operator is the jolliest. He tries to keep the company in good spirits. Pobeshimov has taken over the household duties; he rations the food. Others are sitting near the camp smoking cigarettes and talking. Levanevsky keeps a cool head—he guides them, but realizes that, if they are to be saved, early help is needed."

Another man in Manitoba, Canada, sat in contemplation. "Levanevsky and his companions landed at Longitude 140. The plane is visible on the ice, facing north. All the crew have passed on. They are dead," he said.

In Winnipeg, still another man interested in the lost Russian aviators said he received "from space" this message: "The Russian airplane is badly smashed. A party with it is trying to attract attention. Two others, the most able, are heading southward. Their position is far to the west of their proposed route which was along the 148th West Meridian."

From another part of Canada came another message; "I feel it my duty to let you know where I am sure you will find the lost Russian flyers. I have 'seen' their plane three times—each time in rough ice between two islands, long ones, almost straight north of Winnipeg, a position east of their proposed line of flight."

A note received from a man in California said, "You will never find the lost Russian flyers by searching over the Arctic Ocean. They landed on terra firma. Four are living, and some may now be found to the west of the Great Bear Lake." Another Californian wrote, "The Russian airplane is east of Greenland, four hundred kilometers east of Shannon Island. Some people may not believe this, but 90 per cent of the human animals are morons, thick-skulled, and have no conception of the powers of the human mind."

Before me as I write is this variety of conceptions of the human mind in relation to the whereabouts and condition of Sigismund Levanevsky and his five companions. Obviously, all of the "conceptions" cannot be right. I had laid them aside, and would have paid no further attention to them had it not been for the facts I am about to relate.

In June, 1937, a plane whipped out of Moscow and landed, almost before the world knew that it had started, in the state of Washington. It had flown across the Arctic Ocean, Alaska, and had spent some hours flying over the Rocky Mountains before it could locate its position and find a landing field.

A few weeks later another plane winged its way over the North Pole to cover a record distance, and landed at San Jacinto, California. Both were marvellous flights made by skillful and well-trained Russian crews. The air heroes of the Soviet Union thought they had the North Pole air route taped. The world seemed to think so too.

The first two flights had been made with specially constructed, long-range planes. Then, it seemed to many, it remained necessary to make only one more flight, carrying passengers and cargo in a conventional, multi-motored machine across the Arctic Ocean to put the seal of success and dependability on trans-arctic air routes.

To do this Sigismund Levanevsky and five companions set out from Moscow on August 12, 1937, in a multi-motored plane en route to the United States, intending to make the first stop at Fairbanks, Alaska. Much organization of meteorologists and radio personnel had taken place along the route, and while the weather at Moscow and for the first few hundred miles of the way was not propitious, it was believed that, in general, conditions were as good as might be expected at that time of the year. Levanevsky started in rain, but reported good progress hour by hour.

Nearing the North Pole, he was forced to a high altitude, and above the clouds. He had passed the Pole, and was on the Alaskan side, when he asked to

# I Arrange to Explore the Mind

be given a radio direction of his bearing. Then came the beginning of a message—message number 19. It was partly in code and said, "Motor 34 flying heavily against 100 kilometer wind, losing altitude 6000 meters to 4300 meters 48 . . ."—then came some signals which were listed as "3400" and the code signature "93." Decoded it read : MOTOR ON RIGHT SIDE GIVING TROUBLE. WE ARE FLYING AGAINST A HUNDRED KILOMETER AN HOUR WIND VELOCITY AND HAVE LOST ALTITUDE FROM 6000 METERS TO 4300 METERS . . . WE ARE GOING TO LAND IN . . ."; the jumble which followed was read as "3400." This was not part of the code. The signal was probably an indication of a condition, or a position, but what the latter part of the message was no one living may ever know; 93 was the code signature of Leichencko, one of the crew.

On August 13th the Soviet Embassy in Washington, D.C., issued the following statement: "The plane piloted by Sigismund Levanevsky on flight from Moscow across the North Pole to the United States with an intermediate stop in Alaska is overdue. There is no real anxiety so far for the safety of the crew of the plane. It is thought that a forced landing may have been made between the North Pole and Alaska."

On the 15th, this bulletin was issued: "The Soviet Government commission on organization of trans-polar flights from Moscow to the United States yesterday issued instructions for the search to be undertaken for Sigismund Levanevsky and the crew of the trans-polar plane on August 17th. Moscow reports that various Siberian radio stations heard indistinct signals from an irregularly working radio station, considered likely to be that of Sigismund Levanevsky's airplane. The position could not be established. There is belief that the plane made a forced landing between the 82nd and 83rd parallels, some four hundred and fifty miles from the North Pole on the American side."

On Sunday the 16th, through the agency of Dr. Vilhjalmur Stefansson, then President of the Explorers Club of New York, with whom the Soviet Embassy in Washington was working in close cooperation, I was asked if I would participate in the search. I volunteered my services, and on Tuesday received authority from Moscow to procure for my use a Consolidated Flying Boat known as type P.B.Y. This machine had been the property of Richard Archbold, a member of the Explorers Club. With the help of Major Anthony Fiala and Burt McConnell, an old friend and fellow worker with me on the Stefansson Canadian Arctic Expedition, I was able to assemble from New York all Arctic equipment required while the aircraft was being put in order by Archbold's crew and the O. J. Whitney Aircraft Company.

Air-Commodore Herbert Hollick-Kenyon of Toronto and Alderman Al Cheeseman of Port Arthur, Canada, had responded to my telegraphed request to join me in the search, and to act as pilots while I looked after the navigation of the plane.

On Thursday, thirty-nine hours after I had begun my preparation, we were in the air, flying out of New York toward the Arctic. Within the next ten days we had flown more than ten thousand miles over the Arctic Ocean—more than the equivalent of four trans-arctic flights—yet we had found no trace of Levanevsky. We continued our search in that machine for thirty days, flying a total of over thirty thousand miles. Of the thirty nights we spent twenty-three either flying or sleeping aboard the flying boat; we were only seven nights ashore. Some of the flights were of long duration, the longest being two thousand nine hundred and ninety-two miles over the Arctic Ocean, zigzagging back and forth over the area in which Levanevsky was supposed to be. Still we found no trace of the missing men.

The only indication that they might be still alive was received one day when our radio operator heard on his receiver a signal which might have come from some other operator at close quarters, tuning in on the same wave length. But this could have been a signal emitted from some Soviet airplane conducting a search from the Siberian side of the Pole.

The season for using flying boats in the Arctic came to a close as the temperature lowered to the point at which a skin of ice would soon form over the surface of the lakes and rivers which we were using for our landings. We were then forced to return to lower latitudes.

Although I had volunteered my services, and many governments and individuals had largely contributed toward the effort, the Soviet Government had been put to great expense in carrying out the search. However, unwilling to leave any possibility uninvestigated, and with the hope that the flying activities in connection with the search would prove as valuable as the information Levanevsky might have brought back had he completed his flight to the United States, the Soviet Government asked me to continue the search throughout the winter. To do this would require equipment other than that which I had used throughout the period from August to mid-September, and I began to look about for another machine.

I soon located the best one available. It was the "Good Will" airplane which Mr. Richard Merrill had flown from the United States across the Atlantic to Liverpool and back during the activities in connection with the coronation of King George VI and Queen Elizabeth. It was a Lockheed 10E airplane equipped with extra tanks to give it a range of nearly four thousand miles. It had all available instrument aids to blind flying, and I equipped it with the most up-to-date in radio.

Our search with the Consolidated Flying Boat had been carried out in a period of almost continuous daylight; but the search, if carried on throughout the winter, would have to depend upon moonlight, the light of the stars, and the Aurora Borealis (Northern Lights) as aids to visibility. This, in a measure, complicated the difficulties in connection with navigation. To be effective, every one of our flights would have to be over a distance of more than fifteen hundred miles. If we were to cover all the area allotted to us, we would have to make some flights of more than three thousand miles. To fly from a north-

# I Arrange to Explore the Mind

ern base out over the Arctic Ocean and return—over an area where no known fixed features are marked out on the chart, following a zigzag course, and in a section in which the compass indication in relation to true north varies as much as 100 degrees within a range of three hundred miles—was no easy matter even in summer time when one could expect to be able to get some indication of one's position by taking sextant observations of the sun. It would be much more difficult during winter, when one had to work in a restricted cockpit, behind frosted windows, and in a temperature ranging to 60 degrees below zero, with only the moon or stars to observe as astronomical points of reference.

About four months had passed since Levanevsky had set out on his flight. Many people had given him up as lost. Many thought the proposal to seek the missing flyers by moonlight was fantastic, and they classed those of us who proposed to carry out the search as romantic optimists who believed in miracles. They pointed out the extraordinary danger involved, and believed that it would be impossible to see anything on the ground from an airplane flying high above the Arctic in the depth of winter.

As is often the case, these people based their conclusion upon observations within their own knowledge and personal experience—a habit even civilized man is prone to form. They knew nothing of Arctic winter conditions, nor the history of Arctic rescues. They could not comprehend that the vast and almost totally snow-covered area would act as a splendid reflector for the moonbeams, and that the relatively clear polar air permits more light from the heavenly bodies to reach the Arctic surface than would be the case in the neighborhood of smoke-hazed cities and dusty countries in low latitudes.

The elapsed time since Levanevsky had been last heard from was comparatively short. There was no real reason to believe that if they had landed safely they would not be still alive. Sir John Franklin with his expedition was lost in the Arctic in May 1845, and he did not die on his ice-gripped ship, the *Erebus*, until June 1847. No one from the Franklin Expedition ever returned alive. Yet the search for the missing men kept up for seven years by the British Admiralty, and for two more years by Lady Franklin. The world gained no certain knowledge of the fate of the Franklin Expedition until 1859, fourteen years after it was last heard from.

In 1871 Francis Hall, leader of an Arctic expedition died, and some members of his expedition abandoned the ship which was moored to a floe; in a whale boat they drifted for five months, and for a distance of thirteen hundred miles before they were picked up off the coast of Labrador. They lived on seal meat for much of the time, and had hardly any equipment to stave off the rigors of the Arctic regions.

Levanevsky and his companions set out with full rations for a period of eight weeks. Stretched, these might have lasted twelve weeks; eked out by food obtained by hunting seals, it might have enabled them to live for many years on the Arctic floe ice, drifting with the pack, and, perhaps, covering many hundreds of miles before coming within sight of land.

That is the difficulty, it was said. Wilkins must range over thousands of square miles of snow and ice without even a landmark or indication of the position of the men to guide him. For four months of the time there will be no really bright daylight. Yet, paradoxical as it may seem, that period of the year would, perhaps, be better than any other in which to carry out the search. It was not expected that it would be possible to search every day throughout the dark period. We would restrict our efforts to possibly not more than eight days, or rather eight twenty-four hour nights a month—the four days before full moon and the four days after.

During such time, if the weather were clear, it would be possible, from not too great a height, to see by moonlight a small dark camp silhouetted against the snow.

In the Arctic the moon rises and sets much like the sun in high latitudes. The moon lies completely below the horizon part of the month; the sun part of the year. About the full-moon period there is good light for much of the twenty-four-hour cycle—enough to see objects silhouetted against the snow. It would be easier to detect a flashing light in the moonlight period than it might be to see an object such as a tent on the snow in daylight; and the lost men, if they heard the noise of the motors, during winter, would be sure to flash a light—an action they might not think of carrying out in the sunlighted summer.

There were other considerations. Fog offers great difficulty to the Arctic traveler. Reports based on the experience of Nansen, Sverdrup, Amundsen, and Stefansson, all men who have traveled over the Arctic ice far from shore, indicate that, as a general average, there are twenty-two foggy days in July, seventeen in August, eleven in September, four to five in October, one to three in November, and rarely any from December to March. These estimates are for the Arctic Ocean hundreds of miles from shore, and not for shore areas.

We would, of course, operate from a base on shore; and the fogs there might cause some interference, not only at the start of the flight, but especially when we tried to find our base on the way home. But far from shore, where Levanevsky was supposed to be, we could expect to find a great percentage of the winter weather clear.

However, because of the practical difficulties, restricted cockpits, frosted windows, and the fogging of instruments and lenses due to condensation of the breath, I realized that it would be advisable to use every means known as an aid to my navigation. I proposed to establish two radio direction-indicating stations several hundred miles apart.

The personnel at one station, checking the strength of the airplane's radio signal against that reported at the other station, and interpreting those signals in degrees of arc, could, by triangulation, get some idea of my position at the

# I Arrange to Explore the Mind

time the signal was sent. Then, by means of short-wave radio, they could communicate that position to me with comparative exactitude.

Such radio direction-indicators were not only few in number, but also expensive and difficult to get. I had not succeeded in locating two of them before I was ready to fly north with the machine. So I decided to do without them. This was disappointing, for it meant that if we came down on the ice, and, because of the weather or other difficulties, found it impossible to locate our exact position, others might have the same difficulty in locating us that we were having in our effort to locate Levanevsky.

At about that time I happened to meet Mr. Harold Sherman, whom I had casually known for several years, and with whom, at the City Club of New York, I had sometimes discussed the mystery of the mental processes. I had explained to Mr. Sherman the difficulty I was having in regard to the radio direction-indicators, and he suggested that, since it seemed possible that I would have no radio means of informing the world of my position, we might carry out an experiment.

He told me that for many years he had been interested in the possibility of receiving without written or vocal word, or any ordinary mechanical means, an impression from thoughts of individuals who might be some distance away.

"Working at short distance," he said, "from people in the next room or facing me, and sometimes from people at even a greater distance, I have received impressions which give me an indication of what the other person is doing, and of what he is thinking. It would be great if, when you are in the Arctic, I could receive impressions from you—especially in case you are forced down, and find your radio ineffective."

The idea interested me. For many years, even in my boyhood, I had been conscious of the ability of certain individuals to receive directly within their mind, without mechanical aid, impressions of conditions and events happening far away.

The aborigines of Australia, with whom I associated as a boy, would often give evidence of knowing of some event which was taking place miles beyond their range of sight or hearing. Sometimes, their knowledge of unexpected happenings would be given at such a time as to exclude all possibility that they had heard of them by means of "bush telegraph," or smoke signal—rapid as those means have been found to be.

With an interest in such things early awakened within me, I had often pondered over the possibility of the cultivated, civilized mind, after determined exercise and development along those lines, responding at will to thought forms, thought waves, or thought influences originating in others. I reasoned that if the influence of thought could be directly felt or responded to without the aid of sight, sound, touch, or smell, there need not be any distance limit, if the original thought was emotional, forceful or of sufficient "strength," in connection with the power. I had myself been a party, as have many people, to curious "coincidences"—such as telephoning others who were about to telephone me in respect to a subject which for no logical reason had flashed into my mind.

I have also found my mind occupied with subjects about which no physical experience of mine, or any amount of reading I have done, would have stored up memories which might have emerged from a mental chamber to occupy my attention.

I have often wondered about the possibility of determining the stimuli of such thoughts, and whether they might not be prompted by, or be a "recording of," thoughts which have been emitted "strongly" from the minds of others.

Since some of the so-called "original" subjects with which I have found my mind occupied were of a nature and in relation to things not previously within my knowledge, and not the result of recent research by living persons, it occurred to me that the thoughts of even those who have passed from physical life might be revolving about our sphere, might still be capable of stimulating the thoughts of others, or of being "recorded" by a section of the mental processes of the so called "sensitives."

Through an acquaintance with Sir Arthur Conan Doyle, I had had some little experience with "spirit" demonstrations and clairvoyance, but I was not much interested in either of these phenomena. But I *was* interested in the possibility of dependable and willed thought transference from one living individual to another.

If this quality could be acquired, or developed by individuals specializing in the subject, they, as specialists, might serve a useful purpose in the development of civilization.

All things are relative, and nothing, apparently, whether it be animate or inanimate, can move or be moved without a stimulus or impulse. What was the impulse which had moved me to "original" thoughts, and which had stimulated in the minds of others the impression they had received of realities which were happening at points far beyond the distance from which they may be sensed by means of the ear, eye, or hand?

It might not be illogical to assume that certain parts of the human brain, dormant in most humans but active in some, might be cultivated and developed to a supersensitiveness, and in that condition respond to "thought forms."

Throughout a life filled largely with hard, practical, and physical experiences in great variety and under many unusual conditions, I have had little time to study such possibilities, or to spend in the development of a supersensitiveness within myself of such a brain process—if such a process does exist. But with an accumulation of striking experiences in relation to the possibilities of the mind, and as a result of the fact that many predictions, volunteered to me by "sensitives" each time I am about to set out on an expedition, have proved

# I Arrange to Explore the Mind

correct, I am firmly convinced that the subject may not be put aside as entirely irrational.

In fact, and partly because of circumstances which often keep me many miles away from my friends and without mechanical means of communication, I have often directed my thoughts in the form of a mental message to my friends. I have not enquired of my friends if they have sensed my mental greetings as concrete messages, for I did not believe that such would be possible. But I had faith, and I did believe that my friends would be assisted by my thoughts to keep their memory of me alive. I have been encouraged in the habit by the fact that often when I return from a long absence my friends have said, perhaps without knowing why they said it, "We did not hear from you, but we knew that you were thinking of us." A habit of speech, maybe, but perhaps not merely such.

I could give many records of astonishing predictions fitting exactly the happenings on my expeditions, but I am not particularly interested in those things which savor of clairvoyance or modern prophecy. These are phenomena which were not admitted by those from whom I received my early training—in spite of the fact that these very people were deeply religious, and admitted and accepted the genuineness of the prophecies detailed in the Holy Bible.

I was interested in the possibility of "sensitives" who might respond directly to thought; for if such a quality were to be definitely and genuinely cultivated, that generally unused section of the brain might acquire an ability to serve humanity, and be instrumental in the development of human progress from barbarism toward refinement.

It seems to me that it is unlikely that every person will be capable of developing the same degree of sensitiveness or reaction to thought form or thought waves, any more than every person is capable of developing the same degree of sensitiveness in the perception of tone and color differences, in nervous reaction to touch, or in astuteness in comprehending finenesses in speech and in the written word. But individuals with a conscious reaction to mental stimuli might, with practice and under strictly rational control, aid in the provision of evidence which would eventually throw some enlightenment upon the subject.

So, when Mr. Sherman voiced his opinion as to the possibility of his reacting to mental stimuli which I might induce, I was more than willing to listen to his proposal.

He said, "Do you think we might work along these lines? If it should happen that you come down and find your radio equipment inoperative, will you concentrate on the figures expressing your latitude and longitude? Let us set a time based on Eastern Standard time, say 11:30 P.M. to midnight, on three days of the week when you will consciously and determinedly, with me personally in mind, try to pass on to me your thoughts."

The final arrangement was simple. My task was to search only in west longitudes between 120 and 170 and my travel would take me between latitudes

72 and 87—just five figures. The two immediately following a 1 would be associated with longitude, and that following 7 or 8 would refer to latitude. Another advantage was that lines of longitude in high latitudes are quite close together, and a strip of territory thus designated in five figures would be only a few miles wide and not more than sixty miles long—a comparatively small area to search in case we should be lost and Sherman able to get the message.

Then we began to work out other formulæ, some to be expressed in color. Sherman had had considerable experience, he said, in visualizing both figures and color forms in response to thoughts formed by others nearby. If either my companion or myself were injured I would think of red. If my companion was killed, I would think of black; if both were well, I would think of white. To enable me to fix my attention upon any one of these symbols, I was to imagine that I was looking at the colors as I might see them upon a moving-picture screen. I would imagine that I could see the figures as if written with white chalk upon a blackboard, and vocalize the numbers. The first scene would be two numbers between 72 and 87, the second scene three numbers between 120 and 170, and the third scene red or black or white.

In the event of my being lost, without means of radio communication, the receipt of such information would be of great assistance to anyone searching for me. Even if the result of Sherman's efforts gave an entirely false position and a false indication of condition, it could do no harm; for without any other indication as to where to search, the searchers might just as well look in the position indicated by Sherman as in any other.

I was pleased with the opportunity to carry out the experiment, but hoped that we would not be forced to depend upon it for our succor in case succor was necessary. I planned, if I got into difficulties, to keep a careful record of conditions, surroundings, and of my conscious efforts to get in touch with Sherman by means of thought influence.

Since I am convinced that few human abilities are highly developed without considerable practice, it was arranged that Sherman and I would, in any case and before there was any great emergency, try three times a week to make contact and, if possible, gain some evidence of our ability or inability to communicate. To these efforts I will refer as I proceed with the general account of our winter search for Levanevsky.

# 2

# Enroute to the Far North

ON THE AFTERNOON of October 22, 1937, with Air-Commodore Herbert Hollick-Kenyon piloting, the two of us took off from Floyd Benett Field, New York. We flew through rain and cloud to Cleveland, and there, when the weather showed no signs of letting up, we stopped for the night. The next day we reached Grand Forks North Dakota, early, but hearing that an aerial and civil reception had been prepared for us at Winnipeg, we delayed our arrival until the morning of the 25th.

It was a joy to fly the Lockheed Electra, fitted as it was with all facilities for blind flying—as had been the Consolidated flying boat we had used in the earlier search—but we missed the degree of luxurious comfort we had experienced in the flying boat.

In it we had had room to move about. Two separate compartments afforded space to set up bunks to accommodate all five of the crew. Another compartment of considerable size was available for navigation and radio; the cockpit was roomy, and a compartment in the bow was fitted for the convenience of the observer. We had room to stow all of our supplies and miscellaneous equipment in the rear of the machine, and were thus prepared for a long stay in the machine, away from our base.

In the Electra it was just barely possible, by climbing over the large, extra gasoline tanks, to move from the cockpit to the rear compartment; and there, too, space was limited. It was crowded with emergency equipment and necessary supplies, and there was no room for luxuries. In the Electra the pilot had the same facilities and aids when actually piloting the machine that he had in the flying boat; but once the machine was in flight, there was no chance for him to move from his seat except at the infrequent intervals when the navigator took over the controls and allowed the pilot to scramble over the gas tanks to the rear.

All of the sextant sights of the moon and stars, taken by the navigator, had to be taken through the side windows of the cockpit or through a sliding door above it, which was the usual entrance to the control positions. When this door was open, we were exposed to the elements, and the temperature experienced was the same as that outside of the machine. All calculations had to be worked out with the necessary charts and tables held upon my knees, and the sights for drift had to be made from a cramped position from my seat.

With the Consolidated flying boat we had an economic cruising speed of about 135 m.p.h., but with the Electra we could ramble along comfortably at 160 m.p.h., and step that up to 200 m.p.h., when necessary. The difference of sixty-five miles per hour does not seem much in relation to air speeds, but over a great distance and throughout a days run—as well as to the handling of the machine—it made a great difference.

I realized this particularly when I took over the controls. I had not been accustomed to flying really fast machines and the extreme sensitiveness of the Electra at high speeds was startling at first—although thrilling when one became accustomed to it. When we left New York, the machine was fitted with wheels, and we expected to change to skis at Winnipeg. At Winnipeg no sign of snow was to be seen, and, much to our chagrin, the weather during the fall of 1937 remained unusually mild. Usually in late November at Winnipeg it would have been possible to operate the machine on skis, and quite necessary to use skis at all more northern airports.

Among the crowd that greeted us at the airport at Winnipeg were the Mayor and some of my staff whom I had not as yet met—A. J. L. Dyne, who had been selected as our engineer, and W. R. Wilson, expert radio engineer who had lately been associated with the Marconi Company of Canada. I had sent the airplane skis and radio equipment, which had been arranged for by Reginald Iversen of the *New York Times* radio staff, to Winnipeg by express, and expected that the men would have this equipment ready for transport by air.

At Winnipeg, however, since there was no snow, and it was therefore impossible for us to head directly on to the Arctic, I found the equipment still in crates. As there was no prospect of cold weather for a few days at least, I had time to fly back to Ottawa. There I arranged to obtain two radio direction-indicators which the Canadian Government had most obligingly offered to make available to us.

In making use of this extra radio equipment, we would need the services of another radio expert who would establish his base at Point Barrow, Alaska, some six hundred miles west along the Arctic coast from Aklavik, which was to be our main base of operations.

I had promised Sherman that I would endeavor to keep thought-transference appointments at 11:30 P.M., Eastern Standard time, on Mondays,

# Enroute to the Far North 15

Tuesdays, and Thursdays. I would have to arrange my sittings at various times at night in my location to fit in with Sherman's time.

The first of our "appointments" was interfered with by unpredictable circumstances. At the time arranged for, I was at a dinner given in my honor by the Manitoba League of Aviators, and was in the middle of a speech when the time came to "contact" Sherman.

I could not consciously and concentratedly give my whole attention to review in detail—for Sherman's benefit—the happenings of the day. But I have found it possible, after giving many lectures, to deliver my talk and hold the attention of the public while at the same time occupying some other part of my mind with an entirely different matter.

So, even as I was speaking to the group present at the dinner, I had Sherman and the day's events in mind. I did not know until many days later that on the night of the dinner was to happen the first of an astounding series of "coincidences" in relation to my actual experiences and Sherman's impression of what I had experienced.

Sherman, seated in his room in New York, with his mind blanketed from other thoughts, received impressions which enabled him to write down an almost complete record of the things which had "strongly" engaged my attention throughout the day, and of my actual surroundings and actions at the time he was trying to receive impressions of, or stimulations from my thoughts.

Sherman was in no special position to know just what I might be doing or what I might say. All he knew of my actions was that I was on my way to the Arctic, and he had no means of knowing anything about the conditions I might experience. Yet, some days later, upon receiving his report, I was particularly impressed. Sherman reported on October 25th:

"*Trip satisfactory so far. Equipment not ready. See you break away from others. You will be late in starting. Equipment not complete. Something mechanical not arrived. You are one man short. One man has slight cold. You in company heavy-set man. You have hard time keeping appointment. Impression as if you say 'Wilkins now signing off.'*"

Following this he got the impression that *I left the room to join three men,* then that there were *more than three—quite a group.*

Actually, the detail of what happened, which I had in mind to review or "transfer"—if I had been entirely free at the time—was as follows:

I was pleased with the performance of the Lockheed airplane. I had expected to find equipment which I had forwarded in packing cases to Winnipeg to be uncased and ready for carrying in the airplane, but because the staff realized that, as there was no immediate prospect of snow at Winnipeg, we would have to ship the material farther north, they had left it in the cases. I had talked with Wilson, the radio engineer, and he had strongly advised making every effort to obtain the radio direction-indicators, and, furthermore, he had given me some information which seemed to make that possible. This would involve the services of another man. To obtain this extra equipment would mean some delay. I had heard that Cheeseman, suffering from a cold, could not join the expedition for a few days. I had tried to get out of the appointment for dinner, because I wanted to leave for Montreal, but as arrangements had been made to broadcast part of my speech throughout all Canada and the rest of it over the local station, I had been persuaded to stay. The two men on either side of me at the dinner were short and heavily built, but I had not given this much attention. It was arranged that after I had spoken for thirty minutes I would come to the end of a definite section of my speech. At this point the radio announcer would say, "Wilkins now signing off," and my voice would be cut from the National network, although I would continue to be heard over the local station. This break in my speech naturally caused me some intense thought, and I was mentally, waiting to hear the words, *"Wilkins signing off,"* and wondering if they would come in the middle of a sentence—which might happen if the timing was not perfect.

As soon as I left the banquet hall, I hurried to my own room to keep an appointment I had made with *three men.* They had been in my room for only a few minutes when, much to my annoyance, a *group of about fifteen* uninvited guests arrived, and I could not complete the business with the three I had invited.

I, of course, had no inkling of what Sherman had recorded, and next morning went about the business of the expedition. In order to acquire the radio direction-indicators and to hire the new man, I would leave Winnipeg for Montreal at 2 P.M., that day. The engineers were removing the wheels from the plane, and getting ready to fit the skis, just to make sure that the fit was perfect. Later they would have to remove the skis and replace the wheels.

I had lunch with Kenyon and other friends. Kenyon chose steak for lunch. The fish listed on the menu was of a local variety and, after some discussion as to its quality and its comparison with other lake fish, I ordered it for my lunch. At the table we discussed the matter of shipping our skis and other equipment to some other point, since there was no snow at Winnipeg.

Some thought one place the best, and some another place; there was considerable argument about it. At the same meeting we discussed the new type of ski we would use, and its suitability for our purpose. This subject gave me a degree of concern, for the skis we had were designed to carry a load of ten thousand pounds, while the total load on our plane would be nearer fifteen thousand pounds.

I had, that morning, at the Hudsons Bay store, tried on several pairs of mittens; but finding none suitable, I had arranged to have a pair made of wolf skin, and fashioned in the style preferred by the Royal Northwest Mounted Police. I had purchased a ski cap, with flaps which would protect my ears from the cold.

I joined the plane and flew all night, via Minneapolis and Chicago to New York, and the next morning went on to Montreal. I had hoped that in the plane

# Enroute to the Far North 17

that evening I would be able to keep my "appointment" with Sherman, and was about to do so when an elderly lady sitting next to me ventured to ask me if I thought she might dare to eat something. It was her first flight, she said, and she had been so nervous that she had not eaten lunch or dinner—she was afraid of being ill, but feeling fine and hungry, she wondered if she might eat her sandwiches. She insisted upon keeping up a conversation throughout Sherman's "contact" period.

Sherman's record of his impressions that night, which I received many days later, included among other items:— *"You had fish for dinner. Something in mind about Kenyon—disagreement as to route of flight plan discussed. You seem to be in building with steps leading up to the door."* (The Fort Garry Hotel in which we were staying has a broad and conspicuous flight of steps leading up to the door.) *"Seems as though you have outfitted yourself with a new pair of heavy gloves and a cap."*

Sherman had failed to get the impression that I had left Winnipeg, and that I was distressed at being held in conversation with the elderly lady; but it was not to be expected that he would be able to get everything clearly, or in sequence of the happenings. The surprising thing was that he received the impression of anything at all which could resemble closely anything that had happened that day.

I completed my flight to Montreal on October 28th, and went on to Ottawa to confer with Mr. Howe, Minister of Transport, Mr. J. A. Wilson, Controller of Civil Aviation, and Major Edwards, in charge of the Canadian Signal Corps, concerning the loan of the radio direction-indicators.

All arrangements were satisfactorily completed, and I flew back to Montreal late that afternoon. There I was busy all evening with Mr. R. W. Cook, the new radio man, and the officials of the Marconi Radio Company. They told me that it would be at least a week before I could get the radio equipment. I discussed with them our proposed search for the lost Russians by moonlight, and the possible location of the lost men. I mentioned the experiment I was making with Sherman, and showed them some letters I had received, among them some of the letters I have quoted at the beginning of the first chapter.

Sherman that night received some extraordinary impressions. He recorded: *"You conferred in Ottawa with* THREE *important people. You may be delayed a week or more. Cheeseman is with you—carries a good luck charm. Skis about fitted."*

Sherman knew that I had gone to Ottawa but he would have no means of knowing that I would confer with *three important people* or that I would be delayed because of the radio for a week or more. It seems strange that he should have referred to Cheeseman who was not with me, but who had joined the others of the expedition at Winnipeg that day and here received a "good luck charm" in the form of a wooden penguin which I had carried from Mrs. Burt

McConnell, and which Cheeseman was to receive upon his arrival at Winnipeg.

Sherman's report for that night also includes reference to the Russian fliers. Up to that time I had not received any evidence from Sherman as to his recordings, and was rather skeptical of his meeting with any measure of success in receiving impressions from thoughts expressed by me.

The following day, upon my return to New York, I received the first batch of Sherman's records, but for some time I was too busy to look at them. Even when I had a chance to glance through them, it was not possible for me at once to make a close comparison of the recordings by Sherman with the records in my diary.

But it was obvious that, whatever might be taking place, Sherman was not getting a facsimile of my somewhat interrupted reviewings at the time set for our "appointments." He was, however, picking up impressions of practically *all* of my "strong" or emotional thoughts in relation to the expedition matters, which were expressed at various times of the day. This was of particular interest to me, since my own belief is that a thought strongly ejected will not fade with the first "spread." I believe that it will continue to revolve in our atmosphere or within such bounds that it may act as a stimulus to a responsive mind, and cause some reaction in the mind of another some hours, or even years after the thought has been emitted.

I had then no time to study closely Sherman's report, or further to discuss the matter with him. On Monday, November 1st, I left Newark and flew back to Winnipeg. Kenyon met me at the field to report that the plane would be ready next morning for test flights on wheels; the skis had been fitted and found satisfactory; the wheels would be reinstalled that night.

Kenyon was grieved a little at this, and that no snow was to be found near Winnipeg. He wanted to make the tests on skis, and was impatient, as I was, to get away. But the weather was still very mild in the north; there was no snow on the fields between Edmonton and the Arctic; and at Aklavik, which is on a branch of the Mackenzie River, near the Arctic Ocean, the river was not yet solidly frozen over. It would not be advisable to leave for a week or more.

The Soviet Embassy telephoned me that evening, and I informed them of the progress made with the equipment at Winnipeg, and the possibility of the delay caused by the lack of snow and cold weather in the north.

I went with Kenyon to dine with some friends and met five couples—men with their wives—from Winnipeg's high social strata. There was much animated discussion. Some of the people had met my wife, and I remembered that she had been offered an engagement to sing at Winnipeg, and was inclined to accept the offer. I made a mental notation to tell her not to come. Our hostess was much interested in the psychic. I told her of my experiments with Sherman—which, by the way, I was neglecting by being at the dinner party.

Late in the evening I left the house to return to the hotel where I had an appointment with Miss Kathleen Shackleton—the sister of Sir Ernest Shackleton who had been the commander of the Shackleton-Rowett Expedition which I accompanied as chief of the scientific staff. Miss Shackleton is an artist, and she wanted to do a charcoal drawing of me. Repeatedly, during the drawing, my hostess of the early evening telephoned to know if the portrait was finished, and if I would return to the party. But I could not do so.

Among Sherman's recordings that night is: *"You communicated Russian Government today regarding some flight matter—weather still mild—you advise cannot take off on search under two weeks unless radical change in weather. Kenyon reports plane ready—impatient. Wheels come to my mind—cannot figure connection. You in place near a lot of people. I hear music—talk. Your wife—something you want to tell her. You are talking animatedly with group—several women present. 'Oh yes,' I hear you say to someone, as though you were trying to beg off an engagement, 'Well, not tonight.'"*

On November 2nd, we made the first test flight with Cheesman. He handled the ship well on the first flight, and improved with practice.

We were entertained by the Winnipeg Flying Club for lunch and by the Winnipeg Press Club at dinner. I was a little delayed in getting to the dinner because I was waiting for a telephone call from Chargé d'Affaires Oumansky at the Soviet Embassy, Washington. The night before he had offered to provide me with an extra electric generating set for the Point Barrow wireless station, and that night I was waiting to confirm the availability of the apparatus, the selection and shipping of which would be attended to by Reginald Iversen of the *New York Times* radio staff. It might have been possible for us to have taken advantage of the power at the United States Army Signal station at Point Barrow, but the Soviet Government wished me to be completely and independently equipped, no matter how much the expense. The expense of the extra generating set would be considerable, since it weighed several hundred pounds and would have to be flown over commercial air lines to Point Barrow.

At the Press Club I met Captain Innes Taylor, a man experienced in both the Arctic and the Antarctic. He was with Admiral Byrd at Little America. We had much in common to discuss, for he was interested in ice and polar conditions and in meteorology.

That was election night in New York, but Sherman did not fail to keep his "appointment." Although I was fully occupied at the dinner, Sherman's records for that day included: *"Test flight—in plane—flying over sparsely wooded area—water—scattered houses—skis—you are eager to make test but weather still prevents. You consult naval officer, a man experienced in north—discuss drift ice, ocean depths, air currents. You are awaiting important message from someone in Washington."*

We had actually made the test flight with wheels and that done I was wishing we could make some flights on skis, but the little snow which had fallen the night before had already melted. There was still no evidence of cold weather in the north. I was, of course, still waiting for the call from Washington. I am not sure from which service Captain Innes Taylor received his rank, but he has had much polar experience. We discussed natural science, weather forecasting, and all matters about which polar expeditions are concerned. Sherman had made some "hits" again that night, although he had not recorded some of the most conspicuous happenings of the day.

On November 3rd I had a busy day with the Canadian Customs officials. They were most willing to help in every possible way, and after telephoning to Ottawa I was able to arrange for all of our supplies and equipment to be brought into Canada free of duty. It would be necessary only to check the unexpended portion out again at the termination of the expedition.

On the 4th we prepared the plane for a night flight. At a luncheon given for me by the Searle Grain Company, Limited, I was able to expound some of my theories in connection with long-range seasonal forecasting, and I was pleased to find a great deal of intelligent interest in my work.

As a result of my talk, a vice-president of the Searle Grain Company, Mr. H. G. L. Strange, was moved to write an article, a copy of which he later sent to me. The article was headed with a verse:

>  **What Makes the Rain?**
>  *What is it moulds the life of man?*
>  *The weather;*
>  *What makes some black and others tan?*
>  *The weather:*
>  *What makes Zulus live in trees,*
>  *The Congo natives dress in leaves,*
>  *While others go in furs and freeze?*
>  *The weather.*

I intended to accompany the men to the airport that evening, and to judge for myself whether the conditions were fit for the night flight test. But as I was about to leave the hotel, I received a telephone call from Vancouver, B.C. It was from our radio engineer, R. W. Cooke. His passport was not in order, and he could not be admitted to the United States to carry out a job in Alaska. The Alaska boat was to leave next morning, and after some little difficulty and delay—for it was after government office hours in Washington, D.C.—I was able to get the State Department's permission for Cooke to enter the United States, and the matter was satisfactorily settled. There seemed to be no department or government which would not go out of its way to be of assistance to those of us engaged in the search for the missing aviators.

By the time I reached the flying field, Kenyon, thinking that I might be overlong delayed, had decided to take the machine up for the night flight test.

I had decided from my inspection of the conditions as I drove to the field that the weather was not suitable for a flight. Ice was likely to form, and conditions were too dangerous to warrant us making the first night flight with the machine.

However, before I could reach the plane, it was taxiing out to the runway, and to my dismay I noticed that three women were in the rear compartment. I was furious. The conditions were much too dangerous for any flight, except in emergency, and to take the women up for a "joy ride" under those conditions was risking a little too much. There was no insurance on the equipment or against accidents, and I feared the consequences for myself and for the Soviet Government, should any accident happen.

Shouting, I ran after the plane, but the noise of the engines drowned my voice. The pilot did not hear my shouts, and took off while those of us on the ground at the airport stood by anxiously watching for the plane to circle and land. Several of us with experience in such conditions were sure that the plane would ice up quickly, and, perhaps, have difficulty in getting back to the airport, before it would be forced to the ground.

We heard the engines roaring, and watched the lights of the plane make a slow and hesitating turn. It at once headed back for the runway, and we waited, almost breathless, as it seemed to stagger toward the field. It settled heavily on the runway, bumped, rose, and bumped and bumped and rose three times before it finally rolled with its wheels on the ground.

It was a narrow escape. The airport was not then equipped for night flying; there were no regular, standard lights installed, and there was no de-icing equipment on the wings. The wings and tail surfaces were covered with a layer of ice. It was indeed fortunate that this had been noticed before the machine got far away from the landing field.

This incident made me wonder whether we had made a mistake in not installing de-icers on the wings of the plane. We had thought that once in the Arctic latitudes, where icing was not so likely to take place because of the extremely low temperatures, the de-icing equipment on the propellers would be sufficient to avoid serious trouble.

Sherman's report for that night included:

*"Death of some friend affects you.* (I was surely thinking of death when I saw the plane take off into the ice-laden atmosphere.) *De-icing equipment—need of same. You seem to be in your room—fourth floor."* (The number of my room was 413, fourth floor.) Here was the surprising record—impression of death—de-icing equipment.

Sherman might have at some time or another "imagined" that I would have trouble with ice on the machine, and might well have conjectured that I was concerned about death—or probable death of some of my friends—either those with me or elsewhere, but he had not previously recorded any such impression, and it is not likely that even if he were consciously going through a record of the "possibilities" that he would have hit upon that night as the only night throughout the duration of our preparations and search to record impressions in connection with de-icing, together with the probable death of any one of our party.

On November 5th I had a pleasant visit with the Winnipeg Boys' Model Airplane Club and with many friends, but nothing of great emotional interest happened. The weather in the north was still reported as unusually warm—no snow as yet on the landing fields.

On the 6th I went to the meteorological office to check carefully the weather records, and to compare them with records of previous seasons. Normally, the fields north of Winnipeg should have been frozen and snow-covered by this time of the year, but at only one of our proposed landing fields, Fort Smith, was there any snow. That had recently fallen, and was not expected to remain long on the ground.

The delay was irritating, to say the least, and I wired to Fairbanks and Toronto to ask for a general meteorological forecast for the coming week. I decided that these forecasts would determine whether I should wait at Winnipeg or go on to Edmonton, and from there proceed either to Point Barrow, Alaska, or to Aklavik.

On that day I carefully fitted to the Lockheed Electra machine a new Gatti drift indicator. It is a telescopic affair, constructed somewhat on the principles of a range finder, and is considered fairly accurate when used over sea-level surfaces. It was the first of its kind seen at Winnipeg, and it created much interest among the pilots.

I also fitted a special navigator's prismatic compass in the cockpit. During the early part of the night, we went up for night flying practice, and made several landings with the aid of small and portable flares outlining the field, using only the lights on the machine as a guide for the actual landings. I was able, during these tests, to get some practice with my bubble sextant which I used for observing the altitude of the stars and the moon while in flight.

Sherman's report, made up on the night of the 8th, included: *"Restless. Weather discouraging. Delay again strong impression. Point Barrow. Are you going there for some reason? You turn some sort of an instrument in your hand. Is it a range finder? Now had the phenomena of white lights—like sparks—seemed to appear in the dark."*

Again Sherman had hit upon three conspicuous subjects which had occupied my mind during the period. Should I go to Point Barrow, the range finder (although the instrument was actually a drift indicator), and the spark-like white lights on the field.

The reader should keep in mind the fact that I had no knowledge of what Sherman was getting, and that Sherman had no knowledge obtained by mechanical means of what I was doing or likely to do. It was not until some time after the incidents occurred that I received Sherman's report, and he did not get my comments on his work until some time after that. I had had so

## Enroute to the Far North

many interferences during my attempts to sit quietly on the nights of our "appointments" that I had developed the habit of thinking "strongly" about any incident or event which I thought Sherman might be able to "pick up." Sometimes, when an unexpected incident happened, I would "review" it in my mind immediately, "directing" the thoughts to Sherman at any time of the day when the event happened.

So far, in this writing, I have been restricting myself to incidents and activities which may be verified by the examination of my records written before I received Sherman's recordings.

But these two facts should now be noted. Sherman had already demonstrated his seeming ability to receive impressions of happenings without the necessity of my consciously willing thoughts to him at the time of the "sittings," as well as his ability, on occasions when I was able to keep the appointments, to respond *directly* to my thoughts.

From November 7th to the 19th I made no detailed entries in my diary. However, some of the events referred to in the next few pages may be verified by the examination of the airplane log book and the daily press of Winnipeg, Regina, and Edmonton. It is unfortunate that I did not keep a full record in my diary covering that twelve-day period, for it was a period in which there were many highly emotional experiences. In fact it was because of the frequency of these experiences that I had little time to make lengthy entries. I was either too busy or too tired at night to do much writing.

I did, however, continue, almost automatically, my habit of thinking the unusual incidents strongly "to" Sherman. In a state of high nervous tension, I seemed to have been throwing off thoughts to which Sherman was reacting.

After reviewing the whole of our experiment, I am indeed inclined to believe that Sherman *was* reacting to my experiences. There is no other way of accounting for the number of "hits" which he recorded.

It happened that I received by air mail on November 16th a record from Sherman of his recordings made on November 8th, 9th, and 11th. It was the only time during our experiments that I received his reports a few days after their registration, and I sat down at once, while my memory was still fresh, to comment on the many "hits" he had made.

Most of these, as I mentioned, may be verified by the airplane and engine log books, cash slips from local stores, and the local newspapers.

I had decided to move to Edmonton, and there await snow which would make it possible for us to proceed to Aklavik. On November 9th, we were busy as usual checking equipment, and, since we hoped that we would not have to wait long at Edmonton, I was endeavoring to provide at Winnipeg everything which we might require.

The two three-men tents I had carried on the Consolidated flying boat were not the type I preferred, so I decided to buy canvas and poles—bamboo poles if possible—and make up my own style tent when I arrived at Aklavik where I could borrow a sewing machine from one of my friends at the Hudsons Bay Post.

It was difficult, however, to find such things as light bamboo poles at Winnipeg, and they were not to be had in several stores in which I enquired. The search for the poles took some time, but at last I located some in a small store, and had them packed with canvas ready to be carried in the plane. This was, perhaps, as unexpected an action as might be conceived or conjectured by Sherman, had he been allowing his imagination to play.

Yet when his record for that evening came to hand it included: *"Tent—do you have a portable unit of some kind? Newest equipment in case necessary to camp on ice? I see some equipment packed with* POLES—*to be stored in the plane. Sounding apparatus—were you choosing equipment today?"*

I had enquired, as a matter of fact, in the several sports shops which I visited in search of suitable bamboo poles, about very long lengths of light strong fishing line, such as I have been accustomed to use in sounding the depths of the Arctic Ocean. I wondered whether it might not be advisable to take such apparatus along with me on these search flights. If we were forced down in a storm, and were not sure of our position, a sounding might give me an indication of the nearness of land. I had, therefore, given the matter a great deal of intense thought while visiting the several shops. I could not find the long length of line I wanted, so I finally decided to do without it. Another item in Sherman's record led me to speculate on the difficulty a "sensitive" like Sherman must encounter, on occasion, in "tuning in" on a "planned act" in my consciousness, and in determining whether or not this "planned act" still existed as a thought or had physically been carried out by me.

For instance: On November 6th, I found that I would need considerably more cash than I had with me. It had been arranged to make payments direct to the personal accounts of my staff in their own banks, but owing to the unexpected delay some of the staff wished to receive some payments directly while still in Winnipeg. To make these payments was possible, of course, but it meant that I would have to write my bank and alter existing arrangements. It would also make it necessary for me to ask some bank manager in Winnipeg to cash, on sight, my personal check.

I have an inherent dislike of asking such favors, and since it was a Saturday when the men asked for the money—with Sunday a non-business day—I suggested to them that they might wait until Tuesday for the cash, by which time the bank in Winnipeg could clear my personal check on a New York bank, and provide me with the cash.

This was satisfactory to the men, and I went to the bank to make the necessary arrangements. But like the trusting, good fellow he was, the manager insisted upon honoring my personal check on sight, and I was able to pay the men on Saturday.

# Enroute to the Far North

I had given much concentrated thought to the receipt of cash on *Tuesday the 9th,* and it is curious, although it may not be at all significant, that Sherman in his recordings on Tuesday the 9th wrote: *"You received some money today—I see you get cash—you are in a bank."* On the *6th* I received the cash which I had planned to receive on the *9th.* It was the *only* time I had need to draw cash from a bank during the several months of our work in the north. It was the *only* time that Sherman referred to bank and cash in his recordings.

Could it have been that my subconscious mind had actually given off stimuli on the 9th in connection with a *plan I had arranged to carry out on the 9th,* although the plan had actually and unexpectedly been carried out on the 6th? Or, was it that Sherman's mind had been stimulated on the 9th by thoughts I had emitted on the 6th?

On the 9th some of the men left Winnipeg by train for Edmonton, while Kenyon, Dyne and I stayed over with the intention of flying to Edmonton on the 10th. But the weather on the 10th was bad, and we were delayed until the morning of the 11th—a fact which led to a series of unusual and unexpected happenings.

# 3

## Some Significant Telepathic "Hits"

WE HAD BEEN royally entertained at Winnipeg, and had been too busy to arrange any form of reciprocation before the 10th. But since we were ready to leave, and were delayed until the 11th, I was able to telephone many of my friends and ask them to join us at a farewell breakfast before accompanying us to the airport to see us off.

At breakfast we had a very happy and congenial gathering, and I was moved to give a speech. It was Armistice Day, and since I was supposed to know something about airplanes and their use in war—I learned to fly in 1911, and was with the Australian Flying Corps in the last war—I spoke of the possible use of airplanes in the next war to annihilate civilian population, and stressed what a tragedy it was that we could not confine the use of such modern inventions exclusively to the good of humanity.

I pointed to the spirit of comradeship and goodfellowship which I had observed among all men I have known to be intensely interested in aviation. I regretted that there could not be the same goodfellowship among different nations as there is among fliers of different nationalities—a topic I had often used, but not one which I had discussed with Sherman.

I said, in reply to a question about my association with the so-called Bolsheviks of Russia, that I had found that, in their dealings with me and with others I knew, the members of the government of the Soviet Union had been most gentlemanly; but, I said, the mission upon which I was engaged and the job that I was doing was one that would be expected of any aviator whose experience fitted him to carry out such a mission, the search for a fellow aviator, no matter what that aviator's nationality.

The Mayor of Winnipeg, after responding to my thanks to him for his

# Some Significant Telepathic "Hits"

many kindnesses, said that he was about to present me with a badge of honor, awarded by the city of Winnipeg, and as there was only one lady present—the wife of Captain Innes Taylor—it would give him, the Mayor, pleasure and no doubt add to my pleasure, if Mrs. Innes Taylor pinned the badge of honor to my coat, which she did. I accepted the badge, I said, not only as a personal award, but on behalf of my associates in the search—Air Commodore Herbert Hollick-Kenyon, Alderman Al Cheeseman, and others who were well known in Canada for their courage and competency in carrying out such efforts as we were about to make.

We adjourned from the breakfast room, and as we rode to the airport one of the men with me in the car presented me with a box of cigars. The morning paper, carrying some notice of our expected departure, used a cut of the charcoal likeness which had been drawn by Miss Kathleen Shackleton some days before.

We left Winnipeg amid shouts of good luck and bon voyage from thousands of people. The weather did not promise to be good, and by the time we reached Regina it was threatening ahead.

We were passing over Regina at just about 11 A.M., and we could see a procession of people carrying wreaths to lay at the Cenotaph. In the cockpit beside me were bunches of flowers given me by well-wishers as I left Winnipeg, and the thought occurred to me that it might be a suitable gesture if we were to circle over the Cenotaph and drop those flowers. I gave this some intense thought, but decided that as I was not a Canadian and had no personal association with Regina—had never even visited there—it might be a little presumptuous for me, conspicuously and without invitation, to take part in Regina's Armistice Day celebrations. Because of the looks of the weather, and if the Armistice ceremony had not been in progress, I would have landed at once at Regina airport; but I knew that if I did it would, in a measure, interfere with the ceremony.

Our preparation for the winter search had attracted much notice in Canada, and wherever we went we were shown great official and public courtesy. If we landed without previous notice, some unfortunate official would surely be hurriedly despatched to meet us—and so considerable inconvenience might be caused to the staff of the Lieutenant Governor, the Chief of the Royal Northwest Mounted Police, and the Mayor. For these reasons I decided to try to push on to Edmonton.

We had proceeded less than a hundred miles, however, when the weather conditions were such that it was inadvisable to proceed. We were flying through snow and "icing-up" conditions. The de-icer on the propeller was not working satisfactorily, and the engines, to which the frontal cowlings had not been fitted, were running a little cold.

Finally I decided to return to Regina, land, and have the mechanics put on the nose cowlings, and, if the weather cleared, proceed later to Edmonton. We landed at Regina airport at 12:30, and the mechanics prepared to fit the cowlings.

As I had expected, the airport staff reported our landing, and in a few minutes the aid to the Lieutenant Governor, a representative of the Northwest Mounted Police, and the Mayor arrived. We had, after all, disturbed them, but not until the morning ceremony was over.

The weather between Regina and Edmonton showed no signs of improving, and we were persuaded to stay the night at Regina. We were to have lunch with the Mayor, go to a ball game, have tea with the Lieutenant Governor, and join the Chief of the Northwest Mounted Police at an Armistice Ball to be held in the evening—a very formal affair which was the main social event of the year at Regina.

Our new friends were most congenial, but I could sense at Regina, the residence of the Lieutenant Governor, the headquarters of the Northwest Mounted Police, and of some high-ranking Army officers, a trend toward social formalities. And, as the evening approached, I was considerably worried about my attendance at the Ball. Naturally, flying to the Arctic where I might spend several months away from civilization, and travelling in a plane already loaded with essential aids to the search for the Russian flyers, I had not burdened myself with dress clothes of any kind—and dress clothes, I realized, would be *de rigueur* for the Ball.

So I tried to beg my way out of attending. However, among the friends we had made was a man about my size, and, as he intended to appear in his uniform, he insisted that I accept the loan of his white tie and tails.

It was rather embarrassing for me to accept the offer, since I had never before worn another man's clothes, but to accept seemed to be the only possible thing to do, if I were to be present at the Ball. Unfortunately, no spare dress clothes of Kenyon's size were available, so poor Kenyon had to be relegated to a room on the side, near the ballroom, where I and others of his friends might drop in, and share with him a convivial glass now and then.

I was consciously and considerably embarrassed throughout the whole evening at the thought of wearing another man's clothes, and to this was added further embarrassment due to the fact that while the length of the coat and trousers suited, the waistcoat was a little short in the waist, and insisted upon coming up and exposing about half an inch of the shirt beneath. However, embarrassment in the latter particular wore off as the evening passed, for I noticed that 90 per cent of the middle-aged men's formal clothes showed the same degree of misfit. Many of the men, since the last war, had retired to the Canadian prairies, and wore their full-dress clothing, perhaps, on no other occasion than at this annual ball. In the passing of the years they had filled out considerably, and the most conspicuous thing to me, as the Grand Parade passed around the hall, was the strips of wrinkled shirts showing beneath the strained white waistcoats.

It is interesting to read Sherman's recordings for that evening, November 11th. He wrote:

# Some Significant Telepathic "Hits"

*"In an address these thoughts running through your mind if not actually uttered—aircraft in next war to annihilate civilian population. You make plea for constructive use of airplanes and modern inventions—great forces to be liberated—for the use of humanity—tribute to fellowship aviators of all nations—you are going on expedition to search for Russian fliers in attempt to do for them as you would wish to be done by others were you of a different nationality, down in the Arctic wastes."* These were substantially the words I uttered.

Sherman continued. *"You bemoan loss (lack) of fellowship between nations when so great a work to be done by all."* I had voiced these sentiments, and had told what Captain Frank Hawks had said at the luncheon given at the Bankers Club of New York to the three Russian aviators who flew from Moscow to San Jacinto in July 1937. Speaking of the need for amicable international relations and the difficulty diplomats have in bringing this about, he had continued, "The trouble with diplomats is that they are all a bunch of sourfaces in stiff shirts and high hats who get sore just looking at each other before they get around to talking about peace. Fliers understand each other, they talk the same language, and they'd get along together even at a peace conference."

Sherman continued, *"Someone seems to pin—or put something on the lapel of your coat—either pins a medal or token of some kind. Someone gives you cigar."* The reader will recall the incident of the Mayor of Winnipeg's honor medal being *pinned* on my coat by Mrs. Innes Taylor, and the *box of cigars* received.

Again Sherman: *"You are pleased with charcoal likeness. Tribute to Canadian war dead—flowers dropped from a plane. Something mechanical does not suit—de-icing—more extensive equipment. You in company with men in military attire—some women, evening dress—social occasion—important people present—much conversation. . . .* YOU APPEAR TO BE IN EVENING DRESS YOURSELF.*"*

It seems most remarkable that Sherman should get so many "hits" that night and in relation to a great variety of happenings—some of which of course might have been prompted by his knowledge that it was Armistice Day. But Sherman was not consciously allowing that knowledge to influence him, and the details he recorded dovetailed so accurately with the facts, that it is most unlikely that they are the result of coincidental imaginings.

On the 15th the skis were fitted to the machine and as the meteorological forecast was snow the next day, we hoped soon to make a flight test with the machine on skis. Sherman on that night recorded: *"Skis put back. Get third floor impression as though you so located."* (My room *was* on the third floor of the hotel.)

On the 16th, snow was falling, and we were much cheered by the possibility of a take-off within a day or so. I was checking the personal kits of the men, and found that we would need another pocket compass. I decided to look for a prismatic one which might be of service in case we were forced down, and had to walk home.

I went to the Hudsons Bay store to enquire, and was told that it would be difficult to find such a compass in Edmonton: they could get one in Toronto, but I feared there would not be time for that. The girl clerk telephoned to several other stores, but could not locate the type of instrument I wanted. Then she called the department manager, and he came along with another man to see what could be done. Everyone was most anxious to be of service, and as we were discussing the matter, a third man—a customer—overhearing the conversation, volunteered the information that he had seen a second-hand prismatic compass in the window of a pawn shop, and gave me the address.

I went there and purchased the compass—a military one—which probably had been used by some officer during the last war.

That evening I agreed to see a man who had telephoned me several times asking for an appointment. He said he had an important plan to present to me but did not want to speak about it over the telephone. I had put off most people who had "ideas" they wanted to present, but as this man said he was an old soldier who had served in France, I finally agreed to see him.

When he arrived, he was accompanied by another man, and I was surprised to note that my visitor was blind. He had lost both his eyes during the war, but said he rarely used that fact as a means of obtaining interviews. He represented an insurance company, and reported making a good living by selling insurance.

His proposal was out of the ordinary, and he thought I should be interested in it. It would cost me nothing, although it would require my consent before execution. He proposed to appeal to the public for subscriptions to cover the premiums on a $250,000 policy. It would be a short-term life policy covering the period of my search. If I came back safely, there would be no payments on the policy; but if I did not return or was known to have been killed during the search, the amount of the policy was to go to the establishment of a Wilkins Polar Science fund.

It was a rather ingenius sales plan, and while I doubted the public would subscribe—or that the insurance company would agree to it if they did—I gave the man my permission to go ahead. I came through the search alive, and, of course, have heard nothing further about the matter.

Sherman on this night (16th) recorded: *"You in hardware store for some reason—you buying strands of wire—store you are in seems to have veranda or porch along front; double row of counters; a girl clerk and three men. Artificial flowers—can these have been made by a blind ex-service man? Something made by hands of person handicapped in life seems to have been presented to you."*

Sherman's description of the Hudsons Bay hardware store was correct, and I had dealings with the girl clerk and three men in a matter about which I was giving great thought; but since Sherman described the store and the people, it seems strange that he did not mention the compass with which I was concerned—instead of strands of wire. I may have bought some wire that day, but if so I have no recollection of it. Sherman's reference to the blind ex-service man was, I think, more than a simple coincidence, notwithstanding the fact that he referred to artificial flowers instead of insurance policies. If I may be allowed a little humor, artificial flowers in relation to the proposal might, after all, have been symbolic.

Although the desired snow fell on both the 16th and 17th, the weather was too thick to permit us to leave. I held to the hope of being off on the 18th, and gave orders for the planes to be warmed up in readiness for an early start. We had hired an extra plane to go with us, carrying two of the staff, the radio, and other equipment which would be installed at Aklavik.

Early in the morning of the 18th, the weather—as observed from my room—did not seem favorable, but when the general weather report came in, it indicated good visibility, although overcast weather, at Fort Resolution. It seemed advisable to make an effort to get at least that far, and I hoped that the weather would have improved by the time we reached Fort Resolution, so we could go on still farther.

The 18th was a morning of high tension for me. Parties at Edmonton for the others and for me had become more and more hectic. We were going night and day: the men were feeling the strain, not only of waiting, but of the nearly continuous entertainment.

On the 16th I had given orders that, since snow had come, there were to be no late parties for the crew.

On the night of the 17th most of us retired early. Wilson was feeling rather seedy from a bad cold: he was ready to leave, though not in good health, on the morning of the 18th. We went to the airport.

The weather was certainly not improving, and Kenyon, who is widely experienced in flying the region we were in did not think it advisable to start. The pilot of the hired plane, however, felt we might get as far as Resolution. So I decided to take off, and if we couldn't get through, to return to Edmonton. Any action was preferable to another day's impatient waiting—waiting, and trying to dodge the abundance of hospitality and good cheer of our friends. Not because we did not enjoy both old and new found friends, but because a continuous round of conviviality deprives the body of the rest it should have, if it is to be in the best condition to meet a difficult job in hand.

The snow surface was sticky, and we took almost the length of the field to rise; but the Electra handled well—remarkably well in view of the fact it was loaded almost 50 per cent in excess of the capacity for which it was designed.

Soon, however, I could see that Kenyon had been right. It would be difficult to fly in such weather along the regular course to Resolution, and inadvisable to deviate from that course—which had been selected because it provided possible emergency landing fields. After dodging several snowstorms, we turned back and landed again at the Edmonton airport. Weather in our favor, we would take off the next morning at 9:30.

Not only the morning, but also the entire day of the 18th was replete with incidents to which my thought reactions were "strong." I had Sherman consciously in mind many times, and, by habit, flashed to him a mental review of each incident as it occurred. The evening afforded almost my first opportunity to attempt deliberate and undisturbed contact with him at the hour initially agreed upon.

Although this is not recorded in my diary, I recall distinctly that the wife of one of my friends gave Kenyon and me a dinner that night. Present was a lady, who, our hostess said, was "psychic." In the course of conversation this guest spoke of the "visions" she had seen, and described the attendant condition "as if looking down from a great height to see enacted, prior to actual occurrence, incidents such as fatal accidents" to persons of her acquaintance. She had witnessed (or "seen") her father's death the day before it took place. Since then she has been a little afraid of her power, and tries to take no notice of "visions"—yet they come. Specifically, she reported having had a vision of a mutual friend which purported injury to one of his eyes. (It was not until two years later, when I revisited Edmonton, that I learned the "accident" had happened to our mutual friend. As I did not seek details, it is not possible to verify the facts of her "prediction.")

Anyway, we two went away from the others so that I might quietly review the day's events for Sherman, and later I discussed with her the possibility of discovering the Russians.

Her "special powers" gave no inkling of anything connected with the Levanevsky party.

But here, let us follow through what Sherman received that night: *"I feel you very strongly about your plane tonight—you are either on your way north or leaving very shortly. Snow in the north—more coming—wind. None too good flying weather though ground conditions are better. One of the expedition seems sick. Is it Wilson? Edmonton tries to wine and dine you—you sidestep as often as possible. 9:30 tomorrow you plan some definite action, perhaps resume flight . . . something important set for that time. Wife of friend entertains you at dinner—you show small semi-private group on map where you intend search for Russian fliers. My letter addressed to you at Winnipeg has reached you at Edmonton.* (It had.) *Your location at present in state of change—'under way . . . pushing out,' I seem to hear. Seemed to see your face loom out of darkness suddenly—almost startled me. You appeared to be staring hard at me."*

At the time he was recording, I was actually visualizing Sherman waiting for messages from me. It was almost the first time since our agreement had been made that I had been able to sit undisturbed and review the important incidents of the day.

Sherman made no recordings on regular schedule from the night of the 18th until the night of the 22nd, an interval during which I carried out much important and thrilling work. As was by then my custom, I "reviewed" to Sherman each incident immediately after it happened.

## Some Significant Telepathic "Hits"

We started on the 19th (as Sherman had received the impression) a little before 9:30 A.M. All the crew was on time: in fact, extra early, for the previous morning I had given several of them a severe "dressing down" for being late.

We flew to Fort Resolution—most of the way in vile weather—and managed to arrive safely. This was accomplished largely by Kenyon's skillful piloting, and by sweeping low over the ice-covered river, making hair-raising, steep, banking turns. It required three hours' flying to reach Resolution, and we decided to remain there since the weather was still bad.

Two men in the hired plane, which was forced to land on account of the weather, came in some hours later.

The weather grew worse during the night, and continued bad through the next day. We were compelled to stay at Resolution. Having risked almost too much on the flight from Edmonton, I determined we should avoid any unnecessary danger at this stage since we *must* get to the point of actual search. We would find conditions hazardous enough while carrying out flights over the Arctic Ocean. It was easy to make such a sensible resolve immediately after a day's dirty flying. However, after we had been grounded two days at Resolution, I decided to leave on the 22nd, if possible.

On the morning of the 22nd, we got an early start, and arrived at Aklavik at 1:30 P.M. We passed through some really bad weather on the way, and several times the pilot thought we should turn back. But as our meteorological reports indicated good weather at Aklavik, we continued. The supply plane, however, was forced to return to Resolution.

At last we were at our base of operations, with our equipment all in order. It remained only for us to select the site for our radio direction-finder, and to erect it. Then we would be in position to begin actual nights in search of the missing men. We found conditions at Aklavik most convenient.

Kenyon accepted an invitation to reside at the home of the government doctor with whom he had a long-established friendship. The others of the party and I were to be comfortably established in one of Aklavik's "road houses"— one kept by Mrs. Kost, a veteran of the Arctic who had "manned" her own scow, had loaded it with the necessities for the establishment of a boarding house, and had then floated the scow down the Mackenzie River to Aklavik. She had brought with her two girls of Russian parentage, but who, themselves, had been born in Canada. Mrs. Kost made us most comfortable and provided us with excellent meals.

It was a great comfort to be established in my own room, and to be free from the plethora of entertainment showered upon us in the southern cities. Not that we were to be entirely without entertainment at Aklavik, for at Aklavik there lived the doctor and the nurses at the hospital; a Royal Northwest Mounted Police inspector with his charming wife; several policemen, the missionaries—both Catholic and Presbyterian; a Church of England parson; members of a branch of the Canadian Army Signal Corps; a Hudsons Bay Factor and other traders, besides a number of Eskimo and Indians who mingle at this boundary of northern civilization as they do in few other places. Usually the Eskimo and the Indians, who differ greatly from each other in ethnic traits, do not mix or live near each other.

Surely here it would be possible, when not in flight or actual search, to keep my "appointments" with Sherman with some regularity. Yet, on the night of the 22nd, I was prevented by the arrival of a mounted policeman who came with a message and stayed to talk until almost the end of Sherman's schedule. I had, however, been "reviewing" for Sherman events as they occurred—since the night of the 18th.

Sherman's report of the 22nd includes: *"You following the Mackenzie River in flight—weather, fog, snow—further delay—down at town with old stone fort—Aklavik goal. Radio not perfectly satisfactory as yet.* (This was so. We were having trouble with our radio.) *Another plane following—it takes additional supplies. Now you seem to be thinking consciously and strongly of me for the moment—10 minutes to 12 P.M. Safety factor—no unnecessary risks—no hazards undertaken only on actual flights in search of the Russian fliers—feel some impending action after delay or stopover on flight north of Edmonton. French-Canadian man of French descent talks to you quite excitedly—you have meal with this man—gives you some delicacies to eat. Color white flashes, followed by blue—something having blue color associated with you this day."* (At Fort Resolution I was met by the doctor—a Frenchman who persuaded me to stay at his house. He had studied in Paris, and was most excited to talk with someone who had been in Paris recently, as I had been. His wife prepared a delicious meal, and we celebrated with wine and the usual French delicacies. I was wearing on that flight a new, *bright blue parka,* and was pleased with it—although I did not give it intentionally "strong" thought.)

Sherman had no means of knowing of the delay or stop-over we had made at Fort Resolution. It was intended and assumed that we would make a non-stop flight to Aklavik. He had made many "hits" and the accuracy of his "hits" was arresting. His note that night of my "strong" thoughts directed toward him at *10 minutes to 12 P.M.,* might have been the direct reaction to my thoughts, for it was at about that time that the police officer had left my room. This indicates that "cooperation" was sometimes effective, although Sherman had picked up impressions of incidents of which I had not thought during our "appointment," but which I had broadcast in the interval between appointments.

Communication, by means of our short-wave wireless, was to be much less frequent than I had assumed. During my earlier visits to the Arctic—but from points other than Aklavik—I had been able to communicate freely with the *New York Times* and the Hearst Radio Stations situated near New York. But the winter of 1937–38 was a difficult one for shortwave radio communication all over the world. It was a period of sun-spot activity, and even the most elaborate commercial stations often failed to keep their schedules.

On November 23rd Sherman recorded at the very beginning of his "session": *"First time—don't think it imagination—I get strong, urgent feeling which seems to come from you—eager feeling as though you thinking of me few minutes before time for our attempted communication—figuring difference of our time—getting formulated your mind what thoughts send me. You Aklavik—been there over 24 hours—vibrant quality in the air—different attitude of mind—vital concentrated radiation from you . . . You down to business now—mind freer than it has been for weeks. You seem to be in three-story building—men's rooms down hall—cold and perhaps eight inches of snow on the ground—you make Aklavik in two hops. There seems to be long sloping hill in vicinity of landing field and several cabins on it—or along its base—considerable excitement and interest your doings on part of natives of Aklavik. You've been to dinner with some officials, men, women—several children in teens present."*

What were the facts?

It was the *first* night I had been absolutely alone and able to relax fully and prepare for our "sitting," and carry through with the experiment.

I had an early dinner with the doctor, his wife, and some nurses from the hospital. The doctor's daughter and her Eskimo companion came down from their room to visit us. Wanting to keep the appointment with Sherman, I excused myself and returned to my room, checked the time, and sat waiting for the moment when our "session" was to commence.

I did not write down any account of what I would endeavor to send, although had I known then that Sherman was making so many "hits," I would have been inclined to do so. I reviewed the situation—I did not consciously think of the dinner condition, though it would naturally be fresh in my mind.

I was in a two-storied house. It was really two rooms high, but had a windowed attic which made it look like a house of three stories.

We *had* reached Aklavik in two flights. The village of Aklavik is situated on the sloping banks of a low hill or rise which bounds the river; the river, now frozen over, fronted the houses, and served as a landing field. Eskimos, some of whom I had known years ago when in the North with Stefansson's Expedition, came to visit me and see the plane—and they brought many of their friends with them. We had unloaded the planes. I had been busy checking over the possibilities and advisability of establishing the radio direction-finding outfit at Aklavik, and comparing that site and its conveniences with possibilities at Baillie Island, where there would be fewer conveniences.

Sherman had received all of these conspicuous matters, and had given a good description of Aklavik.

November 24th and 25th were hazy days with occasional fog—a condition which prevented me from taking a trip to Baillie Island to examine conditions there in relation to their suitability for the wireless direction-finder station. I had been rather worried over this delay, and had given the trip "strong" thought. I busied myself making the new tent, planned flights on my maps, and worked out navigation problems in advance to save time when in the air.

On the 24th, three Catholic priests came to dine with me, and they were very much concerned about our search by moonlight. We had a grand dinner prepared by Mrs. Kost, and the Russian girls had made some candy. Nothing of particular moment impressed me on the 25th. My mind was mostly occupied with what we would do when the weather was suitable. I was out for dinner with some friends that night, and had no opportunity to concentrate on Sherman.

Yet, on the night of November 25th Sherman wrote at his "session" a rather lengthy, interesting jumble which included: *"Aklavik—snow—fog certain areas bad flying weather—you are not quite ready for search flights today—going over maps. The supply plane—does it journey to other points to check up on arrangements there?—refuelling."*

(The supply plane, weather-bound as we were, was waiting to take off, and would stop on the way back to check up on the possibility of sending in or carrying in to us additional fuel.)

*"I sense air activity of some sort not related to actual search flight—in preparation therefore, but not in search plane."*

(I had, as a matter of fact, that day discussed with the pilot of the supply plane the possibility of going to Baillie Island, and even of continuing on with that plane to Banks Island, if Baillie Island was not suitable.)

*"Be glad when definitely know you are able consciously cooperate—feel you strongly off and on, but not, as yet, steadily."*

This was in contrast to the two previous "sittings"—when in one instance I was able to cooperate for a few minutes, and in the other after some preparation—and when on each of those times Sherman had specifically recorded his feelings that I was giving him conscious attention. Simultaneous cooperation throughout the period seemed to have helped in a measure, but Sherman, whether I helped or not, was able to get something of the daily happenings.

On Friday, November 26th, the weather was fairly clear, and the temperature was 17 below zero. I had hoped to start fairly early, but the engineers were to have the difficulty usually experienced when preparing for the first time to take off in really cold weather. They started work on the machine at 8:30, and after warming the engines with a stove for an hour, one was started, but the other wouldn't start—an oil line at the back of the engine was still frozen. By the time that engine was warmed, the other had cooled, and the oil had to be drained from both engines and warmed up again.

It was 1 P.M. by the time both engines were running at the same time. By then the skis, which had been pried from the snow earlier, had frozen in solid again, and it was necessary to stop the engines and dig the skis out once more. By the time we actually got started, the weather was beginning to get hazy, and from the air it was difficult to see the ground. We flew over the Mackenzie

# Some Significant Telepathic "Hits" 37

River delta. The islands there lay like strips, and in some places leads or stretches of open water could be seen.

The weather was getting bad, and darkness was setting in. The days are quite short between latitudes 70–72 at that season, so I decided to land near the Hudsons Bay Post at Tuktoyaktok—about midway between Aklavik and Baillie Island. Although we kept a sharp lookout, we saw no sign of the small trading post, but eventually, knowing that we must have passed it, I turned back, and, flying south against the setting sun, we could see the ground more clearly.

We soon located the house, and landed beside it. We stayed there that night, and went on to Baillie Island next day. The conditions for the erection of the radio direction-indicator were not really good. The metal warehouses about the Hudsons Bay Post and the police quarters would interfere with the operation of the machine so I decided to return to Aklavik, and to erect the radio station on the ice on the river opposite our boarding house. This would make conditions much more comfortable for our radio operator, and, being within walking distance of the Canadian Army Signal Corps station, we could use their facilities if required.

We returned to Aklavik on the 27th, and on the 28th erected two tents to accommodate the radio equipment. The next day we erected the aerial poles, and installed the equipment. The Canadian Airways plane had landed at Aklavik during our absence, and it had brought some mail from Edmonton. With it was a letter from Sherman.

On the evening of the 29th, I was able to keep my "appointment" with Sherman, and he recorded:

*"You have received my letter forwarded from Edmonton—are answering."*

(I had answered it, and sent my letter with the Canadian Airways plane.)

*"Had flight feeling—can't feel it search flight yet but some plane activity. Storm area—cold and snow—wind—weather biting—commotion of some sort—motors turning over—crew busy—flight over water and little islands or delta—like strips of land—you come back from flight and remove instrument from plane for adjustment. Padre or priest calls on you. I see you some place where there is an oil lamp—it is carried about and casts a shadow as you move behind or along with the person who holds it. Spirits of men good—though one man—can't sense whether ill or unable be active due his work being temporarily done—he idle, sits around or looks on."*

Sherman could not have known by any commercial or mechanical means that I had received or answered his letter, and he had no grounds for guessing about it. It was not expected that the Canadian plane would arrive with the mail; the trip was a special unscheduled one. He seemed to have got the impression (which I was trying to give) of the difficulties in starting the plane, and setting out on a flight in "biting" weather which froze my cheeks and the noses of the men. He had practically and accurately described the conditions over which we had flown: water and little islands or delta—strips of land. And, perhaps, he had also received an impression about the removal of the instrument. I am not sure of that. A priest, a man who had recently arrived from a long sledge journey along the coast, had come to see me.

One of the incidents of nearly every evening which impressed me was that Mrs. Kost—the proprietor of the "hotel"—would be ready when I came in with an oil lamp which she insisted upon carrying upstairs herself, as she lighted me to my room. It was kindness, attention, and courtesy to the n$^{th}$ degree. On *this* night, however, Mrs. Kost had succeeded in obtaining for me a gasoline vapor lamp which gave a white light, and was much more satisfactory to me than an oil lamp.

It is my custom on expeditions to have my chief pilot do nothing but fly the machine. He gets into the plane after the engineer has warmed up the engines, then gets out and leaves the machine as soon as he lands. This practice was being followed at Aklavik, and there were times when others of the staff were much annoyed by the obtrusive idleness of the pilot. In considering the amount of work to be done the next day, I was thinking of changing my regulations.

On November 30th, we were busy setting up high poles for our radio direction-indicator aerial. They needed to be true north and south. I was determining their position by an observation of the North Star, which was visible between the clouds in the early evening. The engineers were building a hut on the ice near the plane in which to keep their tools, and in which to set a stove for warming the engine oil. Sawed timber was scarce at Aklavik, and I found it necessary to search around to get a few sticks from here and a few from there, having each load hauled to the hut site on sleds. One load was brought on a sled drawn by a chunky, chestnut-brown horse. I had seen the horse in the summer—during our first search—but to see it outside in the winter in such low temperature seemed a little too incongruous to be real. It was about the only time during the expedition that we employed the use of sleds. We worked for a time after dark by the aid of large gasoline vapor lamps, because we were hoping to get ready for radio tests with Iversen of the *New York Times* the next night.

Then we went to a party at the hospital, where we met the matron in charge and several nurses. We were given special cards embellished with sketches, and these cards were used for the collection of autographs. It was a new idea to me, and it interested me particularly since often, after a lecture, I am besieged by autograph hunters who ask me to sign on all sorts of paper—including blank checks when no other paper is handy. (Naturally, I am rather suspicious about these.) But I thought that the supply of stiff cards for a lecture audience might solve the problem of autograph hunters, and enable those who might not want my autograph and who were not pleased with the lecture to say so. Anyway, cards were strongly in my mind.

## Some Significant Telepathic "Hits"

Cheeseman and Dyne went to the school gymnasium, which was a large room beneath the building, and played pingpong with the nurses. There were two pingpong tables among the other conveniences for sport. We had a most enjoyable evening, and went late to bed.

Sherman was much concerned on that night because he thought that his sudden "flash" of *pingpong* might be an intrusion from his recollection that Merrill carried (not in the plane we had, but in another) pingpong balls in the wings of his plane when he flew the Atlantic. He would hardly have guessed we would be playing *pingpong* in the Arctic.

Sherman's report—the night he got the "flash" of pingpong—queried: *"Is there a table in town where people play?"* Additionally he wrote: *"Something hauled on sleds to hangar or building nearby—workmen coming to and fro from plane . . . busy on some job—star reckoning. Are you looking at sky? Searchlight—something by which to scan Arctic wastes when flying and help you pick up landmarks at night—possible evidence of Russian plane."*

Here, apparently, was a case (one of the very few noted in reading through Sherman's records) where his mind seemed to suggest a logical and probable reason why he should have received impressions. He knew that Merrill had carried pingpong balls in a Lockheed plane, and "Light—searchlight" was logical.

I *had* been thinking "strongly" of lights for the light of our powerful gasoline vapor lamps, hung high on poles, shed a strong light like that of a searchlight over the snow, and I wondered whether it might be advisable to keep our airplane landing lights fully on throughout our flights in the moonlight over the Arctic ice. Because of the altitude at which we would have to fly, they would not light up the surface, but they would be a sign for anyone to observe, and would be seen, perhaps, long before the sound of the motors would be heard. Because of this, I think that Sherman "hit" was near enough to be recorded as a bull's-eye. Of course, Sherman included in his records many things which had no place in my experience, so far as my conscious memory provides: for instance, on this same day he recorded, "You thought of Christmas today—remembrance to wife—greetings to few friends."

It is possible these thoughts were in my mind, but if they were I would not have recorded them in my diary, and by the time I received my copy of Sherman's record I'd forgotten whether such thoughts occurred on the day specified. Because of my neglect to keep complete records, I may miss giving Sherman credit for all the "hits" he made. He made so many "hits" that I want to confine this writing to things that can be verified beyond question.

Continuing with the same report (November 30th) Sherman wrote: *"Still feel some special activity Wednesday (tomorrow) work on wireless and tests to be continued and intensified—this one of the most important concerns now."*

I was that night thinking "strongly" about the coming tests arranged for the morrow. We had not yet communicated directly with New York by our own wireless, and I was anxious to get a message through to the Soviet Embassy telling them that our wireless was in order, and that we were ready to start our moonlight search.

From my diary, Wednesday, December 1st: "We had a busy day. The wireless aerials and receiver have been installed; today we installed the engine and generator for the large transmitter. For the first time Wilson was able to establish two-way communication with several Canadian and Alaskan stations, and in the evening about 7:30 we succeeded in contacting Iversen in New York."

I had hoped to be able to send Sherman a real message that night by the wireless operator, but with sending conditions bad, it took Wilson a long time to get through my official message to the Soviet Embassy and a long story to the *Times*. Consequently I omitted the wireless message, and kept my mental "appointment" with Sherman—reviewing the day's events and my plans.

Sherman recorded: *"Real activity now under way—wireless tests about complete. First search flight to take place Sunday or Monday."*

I had been thinking of flying to Point Barrow as soon as our radio equipment was installed—Sunday or Monday, as the case might be.

*"Your wireless operator practicing receiving and sending today—you pick up news from several distant points—communicate with Russian authorities. Name 'Wilson' comes to me . . . he has been upset about something . . . seems to be wireless apparatus."*

*This* day, December 1st, was the *first* day on which it was possible for us to practice both sending and receiving from our ground station, and the message sent that night to the Soviet Embassy read:

"Your wireless station RUPULA is now in operation with the Soviet flag flying above."

I don't know whether Sherman had been aware of the fact that Wilson was the name of our chief radio operator: probably he was not. I was thinking "strongly" of Wilson, and find memorandum in my diary for that day: [radio] "station done in fine time. Wilson good man, fine, steady worker; no hours too long for him."

From December 2nd to 5th, we continued the wireless direction-indicator tests. Many indications of the direction of commercial stations up to eight hundred miles away were more or less correct. One direction-finder would give us an indication of the direction of the station sending a signal, but of course, before we could get any "fix" or indication of the exact location of the station which was sending the signal, we would have to have two radio direction-indicators working at points far apart, and then, by projecting the two directions indicated until they met on the map, we could "fix" the position from which the message was sent.

# 4

# A Fire in Point Barrow

EARLY ON THE MORNING of Sunday, December 6th, our two radio direction-indicators were in perfect order, and we decided to carry out a long-range test flight.

With Cheeseman at the controls we left Aklavik at 10 A.M., and although the weather was foggy and the coastline difficult to see, we flew on various compass courses, testing our apparatus, and arrived at Point Barrow after a five-hour flight. We landed our machine on the lagoon between the two halves of the village. We found our radio operator—Mr. Cooke—comfortably installed. Warrant-Officer Morgan, in charge, had rendered every possible assistance, and everything at Point Barrow was in excellent order. Our electrical generating plant had not yet arrived, and in the meantime we were getting power from the Signal Corps Station.

The settlement of Point Barrow lies on two sides of a saltwater lagoon which is joined to the sea by a short stream. In December both lagoon and stream are covered with ice. The ice on the lagoon was a little rough, but Cooke and Morgan had cleared it somewhat in preparation for our landing.

On the south side of the lagoon, near the edge, was a cluster of Eskimo shacks and tents, many of them close together. Farther off, on the same side, were the wireless station, the church, and the schoolhouse. Behind stretched the bleak, snow-covered Arctic tundra—meeting the sky in an almost flat horizon, relieved only by a few intervening hillocky spots known as *nunataks*.

On the opposite side of the lagoon—to the north—on a steep, bluff-like bank, stood the trading station, several warehouses, and the homes of the various members of the Brower family—owners of the trading station.

Charlie Brower established himself at Point Barrow about fifty years ago, became known as the *King of Northern Alaska,* and rightly deserved the title since he had been magistrate, postmaster, and general director of all official and unofficial matters along the Alaskan Arctic shores for many years. Under his guidance and direction contacts with the native Eskimos were particularly fine.

We had brought with us eiderdown sleeping bags, but we hoped to have the Eskimo women at Point Barrow make for us both sleeping bags and clothes of warm reindeer skin. Since these articles require some time to prepare, I immediately set about finding skins, and engaging the women to make the sleeping bags and the clothes.

With those facts in mind, read what Sherman recorded for the night of December 6th: *"Feel out on the ice with you—test flight, with wireless tested while in the air. You fly about a hundred miles, and return to base at Aklavik."*

(It is possible that this result was prompted by imagination. Naturally, Sherman realized a test flight was likely, and he had previously recorded that we might make a flight Sunday or Monday. He was wrong about the return to Aklavik, but the conditions he described apply to Point Barrow, where we had flown.)

*"See a way prepared in snow for plane to take off. Camp on ice plenty cold—I seem to walk about a block or more from shore line of place where plane located to little settlement on ice—cluster of shacks and tents—they seem to be about eight feet apart, the tents, and, from what I take to be south, you look off to stretches of low, bleak, snow-covered land—hill-like in spots . . . bluff-like hill on Aklavik side with shacks and houses and stores along it."*

(In reality main section of Point Barrow village—not Aklavik.)

*"Do I see sleeping bags—something you crawl into at night—of great warmth?"*

You may notice how Sherman's description of the village fits in with the facts and the conditions. The plane was about two hundred yards from the shacks and tents on one side, and behind them to the north was low, snow-covered country—and on the other side, shacks, houses, and stores. (They were, in point of fact "stores," although I had thought of them as warehouses and a trading station.)

We had been using sleeping bags from the time we reached the Arctic region—bags made of eiderdown, and not particularly warm—but the night to which Sherman refers I was intensely concerned with procurement of *fur* sleeping bags which are of "great warmth." It would be absurd, despite little discrepancies, to relegate all Sherman's record for that night to coincidental imaginings.

On December 7th, Sherman's "hits" were more remarkable, regardless of his continued reference to our location as Aklavik instead of Point Barrow:

*"Seem to see a crackling fire shining out in darkness of Aklavik; get definite fire impression as though house burning—you can see it from your location. I first thought fire on ice near your tent built by one of crew—this may also be true—but impression persists it is white house burning and quite a crowd gathered around*

# A Fire in Point Barrow                                              43

*it—people running or hurrying toward flames—bitter cold—stiff breeze blowing—blaze heightens and throws reflection out over snow and ice."*

It was a dampish, cold, penetrating night with a wind blowing; a night of the kind when hoar frost forms even in below-zero temperature. I was in the radio office at Point Barrow when the fire alarm rang. It was precisely at the time that Sherman should be sitting in his New York office trying to record my thoughts, but, of course, I did not know certainly that he was so doing, as I had no idea he would keep his appointments religiously. I had not been meticulous about my "appointments"; in fact, I was not that night keeping my appointment in the true sense. Instead, I was busy trying to send some wireless messages to New York and Aklavik.

The fire alarm was an extra-long ring on the telephone. There were only four houses with telephones at Point Barrow, and I did not recognize the long ring as a signal for a fire. But Morgan shouted "Fire!"—and together we went to a window, and looked out over the village. An Eskimo house was on fire. Flames were roaring out of the chimney, and as we watched they burst forth and spread to the roof where it was free from snow. Men and women were running with axes and shovels. There was, of course, no water to be had. The men shoveled snow on the fire, and cut away the burning pieces. The fire was soon under control. I talked with Morgan about the frequency of fires, and he said that in the seven years that he had been there fires had been rare. There had been only one of importance, and that was when the hospital and the doctor's residence had burned down a few months before.

We went on with our work on the radio. I was too busy to concentrate for the whole period with Sherman that night, but I remembered thinking to myself, as I worked on the radio, of the possibility of Sherman's getting an impression of the fire.

He *did!*

Coincidence seems unlikely. His report for that night not only gave a good description of the fire, but also included a complete description of our plane, which he had never seen. I question that he had ever seen a picture of it, or had any ordinary knowledge of its general construction. He began with the words, *"Your plane looks like a silvery ghost in the moonlight,"* and ended with the words, *"Plane not yet equipped or stowed with things as it will be on actual hops."*

This was true. I had been giving the plane and its stowage much thought. I had noticed the plane particularly that night, for it was the first time that it had been completely covered with thick rime and hoar frost. The wings, fuselage, and engines were entirely covered, and the plane stood out above the dirtied ice like a spectre.

I had been wondering if it would be possible to include in the stowage a special sled of my own design; one which could be taken in pieces, and put together on the ice, if necessary. I had decided to ask, in the morning, if Dave Brower had in his storehouse the materials from which I could make such a sled.

Sherman's report continues: *"I seem to catch a glimpse of a funeral service—strange sensation this connection . . . the doctor also undertaker—or somehow associated."*

There is no undertaker at Point Barrow; the doctor acts as coroner and records all particulars as to the death. He also performs all the duties of an undertaker up to the point of taking the body to the grave: the Eskimos, themselves, do that.

That evening the doctor had reported to me that a baby had died, and that it was buried during the day. We had discussed at some length the causes of deaths among the Eskimos—mostly tubercular in recent years. We had examined the register of births and deaths at Point Barrow, and had found many curious entries. The doctor was making a copy of the registration of deaths, and pointed out that in many instances the record showed "death from exposure." This, he said, referred to those who had died out in camp, away from the village, and who had been brought to Point Barrow for burial. The doctor at Point Barrow, not having been present at death, and not being required to carry out a post-mortem examination, would list the death as "from exposure," and thus officially end the matter. That method, however, was likely to give statistics a false value, since the persons so designated very probably had died of some disease.

Sherman continued: *"I come upon a dance and social taking place in basement or ground floor of a church or community center—there is a white tower, square, in front. I would say it is a two-and-a-half stories high building. You and crew or members of it have been over and looked in on event."*

All this was quite correct. The Eskimos were having a meeting in the church building, and we had dropped in for a few minutes.

Sherman: *"See some fishing through ice."*

This also was true, for the women had been, *that day,* fishing in front of the village through the ice for a fish known as tom-cod. They often fish—although not every day—through the ice near Point Barrow, but I doubt that Sherman knew it as a result of anything he had heard or read.

He wrote: *"Scrub-like trees. . . several small docks."*

Such things do not exist at Point Barrow, but Sherman also included: *"Some old boats frozen in ice . . . lights in town on high thin poles . . . a road that seems to run a little distance back from the river bank, making a turn around a curvature of land, as river seems to widen out beyond into quite a bay . . . most of houses and stores of any consequence along it."*

That is a rather accurate description of the conditions at Point Barrow—conditions which Sherman had never actually seen, nor had he seen a photograph of Point Barrow.

On December 8th and 9th, we were busy with expedition affairs, and it was at about that time that we were elaborately entertained.

In my diary, I find written for the 8th: "Barrow still keeps up the true hospitality-spirit of the North. We are invited to dinner every night this week.

# A Fire in Point Barrow

First we went to the Morgans', then to Dr. George's, and to the Reverend and Mrs. Klerekuper's tonight. The Klerekupers are the missionaries."

With the Reverend Klerekuper, I discussed Stefansson's book, *My Life with the Eskimo,* in which Stefansson says that if, on his first expedition, he had gone with the expedition boat around the coast, instead of going down the Mackenzie River, he would not have found it necessary to spend many months as the only white man among a number of Eskimos.

Without having lived alone among them, he would not have known more about the Eskimo than the average missionary or trader. Klerekuper said that, before he came to know the Eskimo, he had resented Stefansson's comment, which he considered a reflection upon the missionaries. But since he had come north and had lived among the natives in the established village of Point Barrow, he realized that Stefansson had meant no reflection—he had stated a fact. Klerekuper said that coming as a missionary he found it difficult to get really close to the inner workings of the Eskimo mind, and to discover very much about the purely native customs. He had not worried very long about this, however, since he believed that it was his duty to put before the Eskimos new things and new thoughts, so that if they were so inclined they could adopt them.

We found both Mr. and Mrs. Klerekuper, as well as all the other white population of Point Barrow, charming hosts and delightful people. I visited the school, and gave a talk to the Eskimo children—and to some of the grownups also, including several of my old associates who had served with me on the Stefansson Expedition.

The Eskimos are very clever at interpreting pictures and maps, and I drew for them on a blackboard pictures of my submarine and of the area in which it was used when we were working with it near the North Pole. I also indicated the area over which I would fly in search of the Russian fliers. The Eskimos were much interested, but they were fearful for my safety because they have witnessed so many tragedies as a result of people going out over the ice far from shore.

I find that I noted in my diary for that day that Kenyon and I slept upstairs in our house on separate cots, while Cheeseman and Dyne slept in their sleeping bags on a big bed in a room downstairs. I do not remember "willing" this bit of information to Sherman, but I find in his recordings for the night of December 9th:

*"At Point Barrow you sleep indoors—single cots. At Point Barrow some kind of banquet—yourself and crew guests—seem to see it in church—women of church serve meal—talks? minister. May be coloring, but I see you in connection school—standing front of blackboard . . . chalk in hand—you give short talk, illustrating remarks."*

He also describes the trading-post store correctly: *"Trading-post store is white, almost as wide as it is long, with red roof, neat shelves; canned goods; stove—*

*no, it appears two stoves . . . across from it the trading post appears to be combination saloon and restaurant."*

Across from the store was the restaurant where we and other travellers had their meals, but there are no saloons at Point Barrow. Officially, no liquor is allowed in the area.

There was a good deal which was not quite accurate in Sherman's report for that night. It is possible that his consciousness of known fact may have colored his recordings—as, for instance, when he wrote:

*"Plane in air testing transmitter—impression test flight vicinity Point Barrow, earlier today. Iversen and Aklavik contacting—two-way conversation—hear voices—seem to hear you say, 'Hello Iversen, Wilkins speaking, come in . . . Yes, that's fine . . . I'm getting you all right . . . call numbers SN-2147.' Iversen, New York Times,* calling Aklavik. Two wireless operators—one is with you. Wilson, I believe."

The registration number of my plane was actually N-214 but Sherman's mind had apparently done some coloring, adding the "S" and the "7."

Actually, I had been carrying on a two-way radio conversation between Point Barrow and *Fairbanks,* Alaska. I was speaking to some airplane operators there. The radio operator, Wilson, was at Aklavik and *he* probably contacted Iversen by radio that night. I am not sure, but I doubt that he got through with voice transmissions.

During the period from the 9th to the 13th of December we had some exciting intervals. In addition to a pleasant evening as guests at the home of one or other of the citizens, either white or Eskimo, we had been preparing foundations for our generating set, which we were expecting by freight plane from Fairbanks. I visited the Eskimo workshops where they were making toys and models to send "outside" to friends, or for sale—and I practiced with the radio.

On the evening of the 11th Kenyon, listening for some amateur for a contact, was startled to hear the voice of Pilot Gillam, who was expected to fly the freight plane which would bring in our radio equipment, say (over the radio) that he had been forced down near Cape Halkett—some sixty miles away from Point Barrow. His machine was damaged, but he hoped to repair it, and to fly it in. He needed some material for repairs. He would, he said, have to leave some of our equipment behind, and he warned us against trying to land with our big machine near him—because of the roughness of the surface.

I wanted to get the equipment as soon as possible, so I sent an Eskimo with a dog team to bring it. I knew the country about Cape Halkett well, and I believed that I could find a spot near there where I could land our machine, and then walk to where Gillam was.

The weather was very thick. Visibility was less than a mile. But since it was often better weather at Cape Halkett than at Point Barrow, I decided to take off, and have a look. With Kenyon and Dyne I took off from the lagoon, and, flying about one hundred feet above the surface, tried to follow the coast to

# A Fire in Point Barrow

Halkett. We soon found that impossible; the fog and haze entirely obscured the ground. We had to turn, and experienced considerable difficulty in finding our way back to Point Barrow—where we made a risky, but safe landing.

Warrant-Officer Morgan had an old Ford snowmobile in one of his sheds. It was disassembled, and it had not been in operation for months; but Morgan, always energetic and more than willing to do all possible to help, began to put the snowmobile in order, with the help of our engineers. He thought that even in spite of the heavy fog he would be able to follow the coast line: a rather difficult matter when travelling along northern Alaska—where the tundra meets the sea at low level, and where, in winter, the snow makes it almost impossible to tell the sea from the land.

By 10 P.M. on Monday night, the 13th, the engine of the snowmobile gave a few muffled grunts, and started turning over. By 2 A.M. Morgan, with Ned, an Eskimo, and Mr. Smirnov, set out for Halkett. Smirnov, a young Russian radio engineer who had been at Point Barrow some time, gave able assistance.

He, Morgan, and the Eskimo, Ned, got back the next night about 10 P.M. Or, rather they got within a few hundred yards of Point Barrow when their snow motor gave up the ghost, and definitely refused to move. It had given trouble throughout the trip because of radiator leakage and other things; in fact, the entire trip was more or less of a nightmare for the men. It was dark for the full time, and densely fogged as well.

They had found it impossible to follow the coast, so they had gone out on the sea ice until they came to what is known as *the tide crack*—which is where the moving ice joins the land fast ice. This, usually marked by a line of open water, would appear as a dark strip, and could be something of a guide for the travellers. In progressing beside the *tide crack,* they had got on to some young ice, and could feel the snowmobile sinking beneath them. Morgan had met this experience once before, and that time had gone right through the ice. But this time he knew what to do, and, instead of stopping, he pressed hard on the throttle, and dashed forward at top speed until he reached stronger ice.

He managed to get through, but it was a great risk. If he had stopped, they would have gone through the ice, and probably to the bottom of the sea. By speeding forward, he might have run into open water, but fortunately luck was with him, and they moved on to thicker ice.

Finally Gillam was located. They assisted him with repairs and started back with the load of equipment.

They saw no sign of the Eskimo dog team I had despatched, which did not come in until two days later. Although the Eskimo in charge knew the coast fairly well, he had been lost in the fog, and had wandered about for three days before he could find his way back to Point Barrow. When he finally arrived, his dog team and himself were almost as completely exhausted as was the snowmobile.

Morgan with his snow motor had done the job. We carried our electric generating equipment the last few yards, and started to install it.

Gillam, flying through the fog, came in next day with his stiff-legged airplane. One of the shock absorbers on the landing gear had been completely wrecked, but Gillam had lashed to it an iron crowbar and some blocks of wood in such manner that they served as a strut, and with this makeshift had taken off and landed safely.

It is astounding what those so-called "bush" pilots of Alaska and Canada can do. They fly over thousands of miles of unmapped territory, often alone, and in the worst of weather, if it is necessary to perform a rescue, or to bring in to a hospital some sick person. They sometimes wreck their machines, and leave them, unaided. Some have landed their planes in amazing positions and under almost impossible conditions. And *their* gallant deeds are for the most part all unheralded and unsung. A history of those valiant, courageous, and skillful exploits would make thrilling reading.

What did Sherman report of all these happenings?

On the night of the 13th he wrote: *"Action—seem to feel you in the air—more snow—bitter cold—seems like something has to be soldered—came apart and broke off or cracked—trouble with one of the skis in rough landing due to conditions field and bad flying weather—forced down, either on return flight or test flight along coastline—or out over ocean ice—visibility hazy—considerable wireless activity—air of expectation. 'Abandon search' crosses my mind."*

Not all of these things had happened to me personally or to my plane, but thoughts about all that Sherman mentioned had been "strong" in my mind. Sherman's report coincided amazingly well with the actual conditions. But again, some coloring and association with *known* possibilities comes in to mar the perfection of his registration.

He continued: *"A mind or minds consider further search futile. Russian fliers regarded as dead by certain authorities—think further efforts find them futile—unnecessary risk of other lives—feel these thoughts have been known to you and caused you some concern—other Russian searching parties either called off or stopping soon."*

The very anxious time we had in respect to Gillam's accident and our experience in flight in the fog might have brought conspicuously to the minds of others the extremely hazardous nature of our undertaking. Such was, no doubt, the case with some of the people at Point Barrow who, from the first, had considered our task of a winter search hazardous beyond reason. If Sherman were, in reality, getting indications of thoughts of mine, he might also be getting some influence from the "strong" thoughts of others on the subject with which he was concerned.

He *had* successfully "hit upon" one other of my conspicuous actions that day about which I had some strong thoughts. Because we had found the landing conditions at the lagoon at Barrow somewhat difficult, I wanted to look over the conditions on a larger lagoon farther north. During an interval of fairly clear weather, an Eskimo with his dog team drove me a few miles up the coast

# A Fire in Point Barrow

so that I might examine the conditions. Sherman wrote: *"Seem to see man with dog team take you somewhere on sled."*

On the 14th, 15th, and 16th, we were busy with the installation of our electric generating set, and, once that was installed, we felt that all preparations for any contingencies that we could possibly anticipate had been taken care of. If favorable weather prevailed at the full-moon period, we would set out on our first moonlight search. The weather had not been even fair since we reached Point Barrow.

I had really intended to return to Aklavik, and take off from there; because of delay in getting our electric generating set, however, and since the weather had not been favorable anywhere over the Arctic, I decided to stay at Point Barrow, begin the flight from there, go out over the ice, and return—or at the end of the flight land at Aklavik, after making a triangular flight.

Our meteorologists at Fairbanks, closely studying the weather, said it appeared that it might be favorable on the 16th or 17th. It did clear a little on the 15th, and, not wanting to miss any opportunity, I was half-inclined to go out then, but Kenyon did not feel that we should take the extra risk when—from the meteorological report—it seemed the two days following would afford better conditions. So, we waited.

The weather on the 16th disappointed the forecasters as much as it disappointed us: it suddenly turned for the worse instead of the better, and we were grounded—without option. It grieved me considerably, since I had set my heart on a flight in the December moonlight. But, on the ground, I was able that night to keep my appointment with Sherman. There was little of a striking nature to review accept that Gillam, the pilot, with his patched-up machine, had come in safely; that we were stymied, and that a mail plane from Fairbanks had brought some personal mail.

Sherman's report on the night of the 16th of December included: *"Though this is the day you announced for your take-off, can't feel you this moment in the air; standing by—impression you disturbed account time of moon favorable—definite 'grounded' sensation—temporary stand-still nothing you can do about it. Another pilot, not connected with your expedition . . . can he be mail pilot? See you reading quantity of mail. Triangle impression—are you to fly out from a point or base which represents apex of triangle, then veer right or left as case may be, and come back to original base? I see such lines drawn on a map . . . on next flight you appear to invert the triangle with its straight line base in line with your starting point. This has been given much study."*

I had no thought of "willing" to Sherman anything about the direction of flights. It was my custom to try to "get across" only the things that had happened. Yet, together with the two strong thoughts— *"grounded"* and *"another pilot"*—he had actually picked up what had occupied my mind most of the day. Just as he described it, I was pondering whether to go out from Point Barrow, follow a triangular course which would return us to Barrow, or to fly a triangular course which would take us back to Aklavik. I had marked off both routes on my map, and decided that the weather experienced on the flight plus information reported from Point Barrow and Aklavik (while at the farthest point out) would have to determine the direction of our return.

All of this *obviously* was beyond the possibility of having reached Sherman in New York by any mechanical means. My mind alone had been occupied by this particular problem, and I was at Point Barrow more than three thousand miles away from Sherman in New York.

From the 16th to the 21st of December—the full-moon period—the weather was so bad that there was no possibility of going out over the ice. The delay was certainly irksome—to say the least.

Sherman, on the 20th, recorded the "irksome" symptoms, and a number of incidentals which were correct. Such as, *"Dyne winning a game at cards . . . and you have some rare wine."* The men played cards occasionally, and we had wine at one or another of the homes now and again. That particular night we had some locally made blueberry wine which was very good.

I was considerably upset and disappointed by the impossibility of carrying out a flight during the December moonlight period, and had little that I thought might be "important" enough to impress Sherman. Still he had "picked up" the card game and the blueberry-wine incidents,

In the house in which we lived, a cowling over the stove was set so low that Kenyon, Cheeseman, Dyne, and I had bumped our heads on it any number of times. (It could not bother the owners of the house or the Eskimos, since they were shorter of stature than any of us.) Despite the fact that the bumps were severe and painful—Cheeseman being literally stunned on one occasion—we were slow in developing the instinctive habit of "ducking" to avoid the trouble.

It was most depressing to know that we would have to wait for three weeks before there would be another chance for a flight over the ice. Meanwhile there was little to do except *wait*.

There was nothing in particular to "send" Sherman, yet he recorded: *"I get 'down' feeling—can't throw mood off—it is so heavy—something extra-discouraging has happened.* SUDDEN, SEVERE PAIN *comes to me—right side of head. A number of mental pictures seem to want to form—none comes through strong enough for me to get them. Considerably after midnight, as I have continued to sit in an effort to break through this depressed state, and pull in hazy mental picture."* I certainly was "feeling down," nor was I experiencing any emotional thoughts—except when I bumped my head.

Of course, I was then, as I still am, not *certain* that my "willing" thoughts to Sherman had anything to do with his "conceiving" in his mind what was actually happening. (Perhaps he might have got as many details had he worked independently and without my knowledge.)

But I feel that if you continue an interest in this record you will conclude, as have others, that there must have been some correlation between my thoughts during emotional stress and the stimulation of Sherman's mental

# A Fire in Point Barrow

processes to the point of formulating his impressions into words which described conditions, at times, 100 per cent accurately.

During the interval between the 21st and 23rd (the next "appointment" night) nothing of consequence occurred. I was still depressed because of the failure, through bad weather, to get off during the last moonlight period. There was nothing we could do but wait until mid-January, and we might just as well wait at Point Barrow, where conditions were more comfortable, as at Aklavik.

On the night of December 23rd, I went to see the village Christmas tree (an artificial mail-order one) being decorated and the preparations for the feast to be held on Christmas Eve. At Point Barrow it is customary for everyone, or at least for every family, to provide something for the feast. The arrangement is not compulsory, and no comment is made if someone fails to contribute: it is taken for granted that he has nothing and in any event there will be enough for all.

I went over to a sort of out-building belonging to the church to see the preparations. Many of the delicacies for the party had already arrived, and some of them reeked to high heaven: rancid whale blubber, for instance, is terrific! The Eskimo men were busy scraping it from the slabs of *muktuk*, as whale side is called, to get the cleaner, fresher blubber underneath.

As I sat thinking of Sherman that night, it occurred to me that if his sense of smell were as impressionable as his mind he was surely in for a rare treat. Apparently he was spared olfactory offense, for though his record mentions "depression," the "Christmas tree," and "out of doors—red, blue, and white," it omits reference to effluvium.

Our Christmas passed happily, but was not too bright. It was blizzardy weather. We had Christmas dinner at Warrant-Officer Morgan's. Although quite a number of us were extras, there were presents for all. I received a wonderful pair of *mukluks* (Eskimo fur boots)—just what I needed for the flight. Two frozen fish were presented to me by an old lady who had been associated with the Stefansson Expedition. We had no Christmas tree, so the presents were piled high on a table. We bemoaned the fact that we had not brought with us a real tree from Aklavik.

At Aklavik there are *real* Christmas trees in great abundance and in perfection. It would have been a nice gesture to the Morgans and a great treat for the Eskimos, many of whom had never seen a real tree of any variety. No trees grow in the vicinity of Point Barrow—not even inland. The only member of the tree family the Point Barrow Eskimo knows is a kind of stunted willow which grows only to a height of two or three feet, and that only in some of the inland river valleys.

Sherman, in his record for the evening of December 27th, refers to our Christmas party and to the gifts, and he wrote specifically "NO TREE." He also recorded: *"Red flash tonight as though one of the crew isn't so well . . . (hope this isn't a true impression . . . it crossed consciousness quickly, but I record it in keeping with resolve to make known everything which comes to mind during these sessions.)"*

This flash of red that Sherman saw possibly opens up an interesting angle. One of the radio operators at Point Barrow was quite ill at the time, and I was very worried about him. His illness was often strongly in my thoughts, for if he were not well in time to help in the next moonlight period, we would lose the benefit of our radio direction-indicators.

I did not record in my diary that I had "flashed" a red signal to Sherman, although about this period I was regularly keeping "appointments" with him. Now—as I write—I have no recollection of having thought of red. But we *had* pre-arranged that I would think of red in event of illness or injury. Could it be that Sherman's mind and mine, due to previous strong resolve, automatically reverted to the "red signal" agreed upon? Can it be that some process of the mind—little understood, and not much used—has the facility of "translating" thought to accord with a preconception of the conscious mind?

Some of the most interesting periods I spent at Point Barrow were in conversation with Dr. Levine, then engaged by the United States Medical Service to work among the natives of Point Barrow checking up on prevalent diseases, the vitamin content, and the value of native foods. Our discussion was chiefly concerned with the contrast in habits of the "civilized" and the uncivilized Eskimo.

During the Stefansson Expedition, I spent much time with Eskimos who had seen white man rarely or never before our party came along. They were living in their primitive state, and did not suffer from the many complaints and complications that afflict the Point Barrow natives who have been "civilized" for several generations.

Sherman reports on the night of the 27th: *"Yes, I see you talking to a doctor in Point Barrow with whom you have become quite well acquainted—seems to be a thin, tall individual about fifty years of age."*

This description did not fit Dr. Levine well, but it did describe Dr. George of the Mission Hospital—also a learned man whom I found most interesting, and whom I often had in mind. Dr. George had come north for a short period of rest after several strenuous years of varied practice in the mid-United States. He found his work at Point Barrow interesting—and rather complicated, since the hospital had burned down just before he arrived, and since his patients, regardless of the severity of their illness, had to be "out" patients. There were many suffering from a form of tubercular disease common to the Eskimo—and thought to be carried by the reindeer.

For the 27th Sherman also reported: *"You have watched some fishing through the ice—no . . . it seems like you have tried your hand at it."*

During a stroll over the sea-ice that day (looking for a landing field superior to that of the lagoon ice), I had come across three old women fishing through the *tide crack,* and had sat with them for some time—handling the fish they had caught, but not holding the line. I had reviewed this incident while thinking "to" Sherman.

# A Fire in Point Barrow

The New Year period passed off in the usual way: festivities, parties, and so on, but nothing of special interest to me or to produce "strong" thoughts of the type I believed might be picked up by a "sensitive" mind.

Things were so slack that I did not bother to record anything in my diary until January 8th. I was weary of waiting for the moonlight period to arrive, and a little weary even of keeping "appointments" with Sherman. Neither of us had any means of knowing (at the time) what the other was doing or recording, and frankly, the "sittings" were growing monotonous to me. I am now ashamed to admit my slackening off while Sherman never wavered in his faithfulness.

I did try determinedly to "get through" to Sherman on New Year's Eve, for there was to be a late party in the village. In the fairly early evening at Point Barrow, at a time which corresponded with 11:30 to midnight in New York (due to difference in longitude), I sat in our house by myself. Among other things, "I willed" "Greetings for the New Year" to Sherman, and additionally to many intimate friends—mentioning each person by name. This mental sending of New Year's greetings is a custom in which I have indulged for many years, but throughout the years I have never known whether any friend ever "received" my message except Sherman. He, at his "sitting" that night, recorded:

*"Feel that you are saying 'Happy New Year, Sherman.'"*

It might well have been expected that I would say some such thing, and his recording may have been the result of imagination: on the other hand, Sherman was not familiar with my habit of long years' standing.

Throughout the sittings early in the New Year—rather, *his* "sittings," for I was not altogether faithful—I find Sherman recorded a number of things which may, and quite probably did pass through my mind. He also described incidents which probably happened in relation to me or occurred in my vicinity, although I have no written record whereby to check them, and it would not be wise for me—at this late date—to try to draw upon my memory for corroboration of his "findings" simply because they were *probably* correct.

*Some* of the things he recorded could not be correct, such as: *"flash of chestnut horses pulling sled (conscious mind trying to prevent recording this impression by wonderment if horse this far north), sled seems large—holding supplies and provisions."*

There was no horse at Point Barrow, but there was such a horse at Aklavik, and it had been used to haul our supplies when we were installing the radio on November 20th. It would be interesting to know just why Sherman received this impression on January 4th. (Possibly the earlier incident did recur to my mind on or about January 4th, but I have neither record nor conscious recollection that it did so. *If so,* Sherman might have taken it for granted the "scene" belonged to Point Barrow, where I then was, rather than to Aklavik where I had been.)

Two incidents "recorded" by Sherman on January 6th are interesting, because they are in connection with matters unlikely to be "introduced" by his imagination. He wrote: *"You in company of young man—engineering background—interested in natural resources, particularly oil. You entertained with collection of phonograph records which recall old times."* These things did happen. I discussed with Morgan the oil seepage on the Arctic coast while his small son played some old tunes on a toy phonograph.

For three days before the 9th of January there was blizzardy weather. The wind's direction changed many times, and some irregular crisscrossed snow drifts formed on the lagoon. If they froze hard, they would render the surface unfit for use with a heavily laden plane. This was annoying, for under such conditions we could not use Point Barrow as a starting point—unless more snow fell, and leveled off the surface.

So, I decided to remove most of the gasoline from the tanks of the plane, and, with a slight load, take off and fly back to Aklavik where there would be soft, level-surfaced runways on the sheltered river ways. There the wind could not pack the snow or roll it into ridges. The engines were warmed, and we were ready to start on the morning of the 10th, when a report came from Aklavik stating the weather there was exceedingly bad, and we had to lay over at Point Barrow for *another day.*

On the 11th, the first reports indicated continued bad weather at Aklavik, and I was undecided about making preparations for a start.

Finally, I had about concluded that it might be wise to warm up the engines in the hope that the weather might clear. When a more favorable report did come in, we started the motors. One gave us some trouble at the start. Mr. Klerekuper, desiring to get some motion pictures of our activities, set up several large magnesium flares.

We moved the machine slowly for his benefit, and after going a few feet the skis stuck fast in the snow, and we had to get out and dig them free. During the long spell on the somewhat salty ice the bottoms of the metal skis had become slightly pitted. With the help of the Eskimos—who pushed on the wings—we taxied around in a circle to polish the skis again before attempting to take off. Some of the Eskimos were busy smoothing down bumps on the snow. We finally took off without further trouble, but with the moving pictures and the flares and the taxiing about, there had been much excitement.

I find that Sherman, during his "sittings" on the 10th and 11th, recorded no vivid impression of our having been in actual flight or of our having changed location, yet he did give a very complete description of the happenings at Point Barrow.

He wrote: *"You checking conditions—hear motors turning over—distinctly see you and crew running across snow field in a hurry; lights on field—men gathered—late weather reports more favorable—attitude of men tonight; feel them acting as a group. Many people moving about—unusual activity at Point Barrow. Eskimo men and women, heavily garbed, out watching—one motor acting up—takes long to warm up—seem to see plane move along runway on skis then, with men help-*

# A Fire in Point Barrow

*ing, return to starting point after some difficulty—motor turning over for taxiing purpose—checking plane's traction and runway conditions—fixing certain uneven spots—packing them down and filling in—definite action preparatory to take off."*

To this point Sherman's "recording" is an almost *exact* and nearly complete description of the actual happening, yet he *concluded:* "*You* MIGHT *take off after all."* We had taken off and had landed at Aklavik by the time he was "sitting."

It is such miss-hits which make the study of mental phenomena so puzzling, and the interpretation of our experiences so difficult. I have become more and more inclined toward the belief that it is the highly emotional or exciting thoughts that are most apt to get through, and to make an impression on another mind.

In this case, for instance, I was deeply concerned about the weather. I had hurried the crew out to prepare the machine which we unexpectedly found was nearly covered with rime and hoar frost. This was annoying, for we had thoroughly cleaned it only the day before.

Mr. Klerekuper had set up lights and prepared brilliant magnesium flares to be used in taking his moving pictures; an unusual number of people were about to see us take off; one engine failed at first to start; the skis stuck and the men helped us to clean them by pushing on the plane as we taxied slowly—all extraordinary actions which had not been done before. It was all exciting to me, and Sherman "recorded" it.

But the *actual take-off* was *not* exciting, nor was the flight. *These* were *ordinary* things *for me* and although Sherman received the "impression" of all the rest he did not "get" the thought of what was, to me, a casual performance. *Had he been guessing,* the take-off and flight would, presumably, have appeared the more important things.

We experienced considerable difficulty with hoar frost forming on our windows as we flew from Point Barrow to Aklavik—despite the fact that we had installed the best window-protection device then available, the one used by the majority of Canadian pilots who fly during winter. I, therefore, decided to try a new device, which I installed with the help of Wilson and Dyne.

On the 13th, I was able to sit and "think" things to Sherman, but the only indication that he received my "impressions" was that in his recording he mentioned, *"Trouble with wireless sending. Something has been found impractical and has to be changed—you supervising work. Am impressed with January 15th as your probable take-off date."*

We were having great difficulty—as were radio operators all over the world—in getting our messages through, and it was because of this that I was not sending messages by radio to Sherman, reporting on our experiments. I was hoping to get off on the 15th, and had worked out my navigation problems in advance for a fixed-time take-off on that date.

On the 14th, however, the weather cleared, and we had a report from our meteorological forecasters that the "night" of the 14th and the whole of the

15th might be suitable for a flight. I had completely stowed the plane with all our requirements, and, when testing our instruments, found that there was some trouble with our airplane wireless transmitter. But Wilson, by working long hours in extremely cold weather and under difficult conditions, managed to get it in order.

At 5:50 P.M., on January 14th, we taxied down the river, and at 6 P.M. we were in the air on our first search by moonlight. The moon shone bright and clear, but we were still having trouble with the hoar frost on the windows, and repeatedly had to shave it off with a razor blade attached to a handle. When it was necessary to get an unobstructed view from the plane, we had to open the windows, and endure the blast of cold wind in our faces.

We soon came to scattered clouds beneath which we flew. When covered by the shadow of the clouds, the ice surface was not clear, and the rough patches were difficult to distinguish from the smooth; but in the clear moonlight the rough ice threw heavy shadows, and the snow drifts were plainly seen. It would have been possible from five thousand feet (the height at which we were flying) to see men *moving* on the ice, even had they been a mile away. If, however, they were standing still, their shadows and forms might not have contrasted clearly with the forms and shadows of the rough ice, and it may not have been possible for us to distinguish them.

A similar condition would have existed had our light been from the sun, so we felt we were in no great measure handicapped by the fact that we were searching by moonlight. We could, however, have seen for a greater distance by sunlight, and so have envisioned a greater square mileage as a result of the flight.

After flying northward about four hours, we came to more heavily clouded areas, and came down to an altitude of two thousand feet. But since we could not see the surface more clearly than we had at a greater height, we again climbed—and soon after we were in snow-laden clouds. The hard snow crystals coming through the open window beat against my face until it felt raw and bruised. It was useless to go on in such weather, and, at 10:45, I decided to turn back.

On our homeward journey we found the conditions through which we had passed had changed. A high wind had sprung up, clouds had gathered over all of the route, and I could not get any sight of the moon or stars. Nevertheless, our navigation, with the help of observing our drift in relation to the darklooking strips of open water which divided the floes of ice, was fairly accurate, and we landed safely at Aklavik at 3:30 A.M.

Nine and a half hours of flying had taken us to Latitude 77.30 N. and Longitude 137 W. and back, and we had covered a distance of approximately fifteen hundred miles. We had been forced to return before we reached the area in which the Russians were thought to have landed. But, since it was then six months after they were reported lost, they were as apt to have drifted into the area we did cover as to any other. I had kept a close watch on the ice until it

# A Fire in Point Barrow

became obscured by haze and snowstorm, and had seen no object resembling a plane, a camp, or a flashing signal which might have been sent by men had we passed near them.

We refueled the machine immediately to be ready for another attempt, but the weather still held cloudy. It remained cloudy and unsatisfactory throughout the 15th, and on the 16th the conditions—as judged by our local observations and the forecasts of the experts at Fairbanks (who were getting their reports from all sides of the Arctic Ocean)—were impossible for effective search. The 17th, too, the weather man indicated, would show low clouds over the Arctic Ocean, and fog along the coast—making for poor visibility in which it would be useless to fly.

What did Sherman get in relation to the search flight and the conditions? It so happened that the flight came between the days of our "appointments," and in the meantime Sherman had read in the newspapers my report of our flight, and a full description of the conditions. For this reason, it was not possible for him to sit with an unbiased mind to collect anything I might have reviewed for him.

On the 17th he recorded: *"Of course I know of your search flight on January 15th. You have overhauled your plane and seem to be thinking strongly of another attempt tomorrow"* (which was, of course, likely, but Sherman had no means of knowing whether or not we had actually taken off on another flight as intended, and expected). *You have experimented a little with ESP cards."*

With the mail I had received upon my return to Aklavik was a set of the so-called ESP cards—such as were used by Dr. Rhine in his experiments in "extrasensory perception." I was not particularly interested in carrying out experiments with these cards, because it seemed to me that, if there were one especially difficult way to demonstrate the possibility of thought transference, it would be with five marked cards.

I had tried, and I found it impossible for me to pick up one card, think of it intently for a few seconds—or minutes—and then, as I looked at the next card with a different form, completely dismiss the impression of the preceding card and its markings from my mind. There could be nothing emotional or gratifying about looking intently at a card, and the very fact that one was "determined" to forget the first card and remember only the second, appeared to keep both cards in mind. Anyone, assuming he could receive an impression of my thoughts, would have received a very confused notion and been forced to make a snap guess as to which I was really trying to think about. I had made no effort up to this time to use them in relation to my appointments with Sherman.

On the 18th at Aklavik, it seemed as if we might have another chance for a flight, but the meteorologists reported that conditions were still unfavorable over the Arctic Ocean. This brought us about to the end of the January moonlight period, and we mournfully looked forward to another long wait before it would be advisable to make another attempt.

Sherman recorded, among other things, that night: *"You expected earlier to take off on a second flight . . . and then decided against it—possibility now of having to wait until next month for further flights."*

Aside from one disquieting discovery, there was not much activity of importance between the 18th and the 24th. We found, upon uncovering the engine and propeller (which had been shrouded in the dark immediately after we had returned from our flight on the 15th) that one end of the propeller was sharply bent at the tip. Evidently it had struck some object as we were taxiing, and in its condition must have exerted a very harmful influence on the engine throughout our long flight.

I recalled that when taxiing we had heard a sharp report, or noise, and I'd said to Kenyon, "What was that?" He had replied, "Just a backfire from a cylinder, I think." Throughout the flight we noticed the throttle lever used to synchronize the engine revolutions was slightly out of line, but I thought it due perhaps to the effect of cold, or that possibly one of the engines was not tuned up quite right. Memory of the supposed backfire encouraged this explanation, and I gave the matter very little thought—forgetting it entirely by the time we returned to our base. The engine had worked throughout the flight without any vibration of which we were conscious, but now that the damage was noticed we wondered if it had really caused some trouble in the bearings. Dyne straightened the propeller and started up the engine. It seemed to be in good order, and we dismissed the matter from our minds.

On January 24, my diary records: "A little sunshine this morning. Clear. Temperature 40 below zero and light northerly wind." It had been clear on the 23rd also, but since in the higher latitudes the sun would still be below the horizon, and as there was no moon, it would not be profitable to fly out over the ocean in search.

I went for a walk along the river. The sun had set, the stars were brilliantly clear overhead, and as I walked I was thinking of Sherman.

Sherman's report for that night—January 24—includes, *"It seems to be clear tonight for a change, but too late now for you to undertake search flight until next month. You observe heavens and contemplate mystery of space . . . are you looking at sky and thinking of me? I seem to get this impression."*

The words in my diary are: "Walked along river bank. Saw ptarmigan track and one lemming track—nothing alive—(thinking of Sherman)."

Sherman's report continued: *"Thoughts go to stars in the sky and more definite sensing your conscious thought in my direction. You seem warmly clothed—see face clearly—hooded garment—fringe of fur about face—heavily lined head covering—outdoor apparel—thick coat coming a little below waist, baggy, heavy trousers tapering to ankles, heavy boots, combination moccasin-galosh (type hard to describe)—you are out somewhere. See plane clearly—not your plane. You talk to mail pilot."*

The Edmonton mail pilot left that day, and I was at the plane to see him

# A Fire in Point Barrow

take off. The description of my dress is correct and complete. The boots were of the type known as *arctics*—soft rubber bottoms, felt tops.

On January 27th, Sherman recorded among other things: *"A dog seems to have been injured at Aklavik and had to be shot—quite a strong feeling here."*

I have no record of the incident in my diary and would not swear as to the actual day, but I well remember that about this time, as I was walking along the river, I came upon the body of a dead dog: it had been shot through the head. I thought about this strongly for some time, and wondered the reason for the killing.

In Sherman's report for January 31, I find, *"You seem to have batch of mail ready to send to States. Seem to hear you say to someone: 'If radio communication keeps on this bad I'll have to develop telepathic communication with Sherman in order to get anything through.'"*

My diary records, January 30: "Answering letters. Thinking of Sherman."

On the 31st, we were trying desperately to get through by short-wave wireless to Iversen in New York. We could just hear him, and thought he could hear us. We sent the message, but we got no acknowledgment, and could not contact him again. I was most anxious to get the message off that night, and said to Wilson almost the very words Sherman recorded . . . *"If radio communication . . . etc."* One of the messages I wanted to send was to Mr. Oumansky at the Soviet Embassy at Washington.

He had sent me a message stating that he had learned that the son of Mr. Smirnov—the Soviet representative at Point Barrow—had died. Smirnov's wife was extremely ill, worrying, and longing for his return. (He had left Moscow soon after marriage, and had never seen his son.) The Soviet Government wished to relieve Smirnov from duty at Point Barrow so that he might return to his wife, and, if I could get along without his help, they would make arrangements for him to fly from Point Barrow to New York, and take an early boat for Europe.

I was tremendously impressed by this attitude of the Soviet officials. I knew, of course, that Russians are normal people, with all the fine human qualities possessed by any other nationality, but I wondered if the government officials of other countries would be so quick to relieve an important officer from duty, provide a special air transportation for him for the first half of a five-thousand-mile journey, and a quick passage home so that he might console a wife who was in an extremely upset mental and physical state. Naturally, no matter how important to me, I was willing to get along without the valuable services of Smirnov, and was anxious to get my approval into the hands of the Chargé d'Affaires at the Soviet Embassy at Washington. And the extremely bad radio conditions were holding up communications.

On February 1st, we tried to get Iversen, and could sometimes faintly hear him. We knew he would be trying—for he had not received our messages from the previous day—but we could not get a readable message. I went out with my cameras to take some pictures, and filmed a dog team coming in from a long trip; some children with their sled; snow on trees, and so forth.

Sherman reported that night: *"Need for certain kind of tubing comes to mind."*

I had that day searched for and found, after looking for a long time, a piece of steel tubing which I fashioned into a skin scraper, and added to my kit. It would be used to keep our skin clothes in order if we were forced to use them on a walk home from an emergency landing.

Sherman continued, *"See a dog team and a man on snowshoes—looks as though he is wearing a thick, black stocking cap with separate ear muffs that fasten with band over head ... rough growth of beard ... you talk to such a man ... dogs huddle around sled ... seems filled with furs and hides ... light snow falling, cold."*

This very well describes the man I had filmed, and I was impressed with his stocking-like head covering over which he wore a regular Arctic type cap with ear flaps tied over the top. I do not remember, however, that I included during my "session" with Sherman any particular reference to the man, although I did think of the picture-making. It might be said (if we agree that there was any real connection between what had impressed me and the impressions Sherman got) that he "picked up" the thoughts "cast off" as I spoke with the man, rather than the "cold" impressions I tried to emit during my "appointment."

On February 3rd, I received word that Pilot Gillam, with whom we had had contact at Point Barrow, had experienced another accident, and was down out of fuel somewhere near the mouth of the Chipp River. His plane was undamaged, but he had been lost in a fog and had run out of gas. I wired Mr. Beliakov, the Soviet representative at Fairbanks, that I would take off early in the morning, drop some gas for Gillam at Chipp River, and then continue on in search over the Endicott Mountains for the Russian flyers.

It was the opinion of some people that Levanevsky had reached the Alaskan shore, and had crashed into the mountains. The Chipp River is in a neighborhood which had not been thoroughly searched.

On that day I had received a message from Mr. Oumansky, asking me to write an article to be wirelessed to Russia with reference to the Soviet Polar party which was about that time reported to be in difficulties because of the breaking-up of the ice floe on which their camp was established.

Sherman, although he recorded, *"Flight of Russian polar party being watched with interest,"* seemed to have got no impression of the preparations I was making to go out in search of Gillam next day. He did, however, record, *"You charted course of proposed search flight, and have gone over plans carefully with crew—your next flight seems to be figured roughly as consuming between 1000 and 1200 miles (out and back) with alternate in your mind of 900 miles' flight."*

This indicates that Sherman was getting some hazy impression of what was

# A Fire in Point Barrow 61

in my mind. Our flight to Gillam and along the Endicott Mountains would not be more than twelve hundred miles since we would have to work in the relatively short period of daylight. All our flights over the *Arctic Ocean* had to be planned for a distance of more than two thousand miles, if we were to cover the area where Levanevsky had been when he sent his last message. I had with Kenyon gone over the plans for the trip to aid Gillam, and for the proposed mountain-area search. I would not have done this in respect to the Arctic Ocean flights, because the actual course on those would depend upon the weather reports we received in relation to probable clear spots over the ocean, as forecast by our meteorologists.

So I think we may clearly assume that, while Sherman was not getting a precise account of my thoughts, he *was* reacting to the thoughts of the day.

On Friday, February 4th, I was up early to get the weather report, hoping that we might be able to take off, and drop the gasoline to Gillam. The wind was strong at Aklavik, and there was a considerable drift of snow along the river surface. Radio conditions were bad, and although our schedule with Point Barrow was at 7:30 A.M., we could not get through until 8:10. Even then, the signals were so bad that we could not get a comprehensible message.

Meanwhile the wind was rising, and I went to the plane to stop the men from starting the engines. One of the engines was, however, started by the time I arrived, but the weather was so thick that it would have been useless to have gone up in flight. At 9:30 we received a message from Barrow. It had been relayed through Fort Norman. It said that Gillam had prepared a landing field for us on a sandpit, and that the dog teams sent to his aid had not arrived. We could do nothing to help Gillam that day; the weather was too bad, but we made arrangements to contact Barrow that night.

We got through to Barrow at 11 P.M. Conditions were bad, but we managed to receive a message saying that Gillam had contacted them earlier in the evening. The dog teams had reached him with a supply of gasoline, and he expected to fly to Barrow the next morning. There would be no need for us to fly to his aid.

On that night during my appointment with Sherman I tried, for the first time, to concentrate on the E.S.P. cards, and turned up the cards in the following order: star, cross, circle, square, wave. I found, as I looked at the second card, and then the third, and so on, that it was difficult completely to blanket my mind from the mental impression of the previous card. As I turned them over, it seemed to me that it would be surprising if Sherman recorded anything like the sequence I had arranged. Sherman's record shows that he had that night written down: *circle, star, cross, circle, wave.* He had got the star, cross, and circle sequence right, but he had missed the square entirely in his first recording, and then, shortly after, seemed to see cross, circle, square—another sequence that corresponded to my list. But altogether the pick-up was by no means extraordinary.

Sherman does, however, record, *"Impressions of take-off tomorrow morning... in search for Gillam... some wind...."* Sherman knew from press reports that I was hoping to take off in search of Gillam, and he might have guessed that it would be soon; but he had no real knowledge of our plans.

He also included, *"You appear to have been heavily dressed at the time of the experiment... and now go outside again... some work still to do."* I left the appointment with Sherman to go out to the radio shack on the ice on the river, and it was some two hours after the appointment with Sherman that I received the report that I would not have to go in search of Gillam.

I did hope, however, that I would be able to make the search flight over the Alaskan mountains, but on the Saturday the weather was again not good.

Sherman records on the night of the seventh, *"Could not feel you in flight on Saturday, and sensed that Gillam had gotten through to Barrow without assistance from you... the tail of the plane... something about the rudder control... radio activity... a hand tapping at key sending messages... feel you have some communication with Iversen tonight... something fixed today with a ladder of some sort... see one of the crew reaching up."* Sherman was right about Gillam and about the other items as well. I had carefully studied the tail of the plane that day to see whether we might use that for making a sled—in case we were forced down, and had to walk home. I decided that it would do, and because of it we would be able to leave the wooden sled, made at Point Barrow, behind, thus saving weight and space.

Sherman had made a "hit" also in respect to the radio: *"a hand tapping at the key,"* for I find a record that I *personally* and for the *first* time since arriving at Aklavik, had tapped out the message to *Iversen*. I seldom handled the key myself, and the personal effort of that night must have been given some thought—although I do not remember having willed that thought to Sherman.

The ladder incident mentioned was also a "hit." The men were erecting a shelter over one of the engines, and I was much concerned about the stability of the ladder on which they were working. In fact I have a fear complex in relation to ladders, having fallen from them and with them several times, and have three times sprained my wrist as a result of falls from ladders.

Another of Sherman's notes was: *"Plane taking mail from Aklavik."* This was true. I was aboard that plane, but Sherman did not register that fact; and there were several other important matters that Sherman did not report.

On the 6th, we had intended to take off for the flight over the Alaskan mountains. The engines had been warmed up, and were idling when Kenyon and I got into the cockpit. Then, one of the engines stopped; the revolutions gradually slowed down, and eventually the engineers found that some of the bearings had seized. The trouble seemed to be with the main bearing or the bearing of the master crank rod. Dyne reported that the oil pressure at the gauge had been correct throughout the warming period, and he said that the only reason he could suggest for the seizure of the bearings was that the engine must have been damaged during our flight over the ice.

# A Fire in Point Barrow

We had flown, it might be remembered, with one end of a propeller slightly bent. Since that time we had straightened the propeller, and we had run the engine for an hour or more just to see whether any serious damage had been caused during the flight. Throughout that test the engine had behaved in a normal manner. It was lucky for us that the rather long test had been made on the ground, for that test had brought the engine to the point of breakdown just a few minutes before we were about to take off.

Had the engine seized as we were taking off, it would have meant disaster. It was fortunate for us, also, that the mail plane from Edmonton arrived that day, and I was able to decide, after some communication with the Soviet Embassy, to go out with the mail plane, and see about getting a new engine. The one we had trouble with would have to be returned to the factory for an overhaul before it would be safe to take out on long flights over the ice.

So, on the 7th, Monday, I started with Pilot McMullen in the mail plane for Edmonton. We had trouble with McMullen's engine on the way—some trouble with the oil feed—but it did not cause a forced landing. We were able to continue until we came to a regular stopping place, and there the trouble was remedied. We stayed at Fort Simpson for the night.

The temperature was very low at Simpson on the 8th, and the struts of McMullen's plane were frozen—one up and one down. The mechanics set fire to the oil about the engine while warming it; but the fire was soon extinguished, and it did not cause any damage. With the plane lop-sided because of the frozen struts, McMullen, with great skill, took off, and we flew on to Fort McMurray. We met bad weather toward the end of the day's flight, and learned that the weather was bad at Edmonton, so we stayed at Fort McMurray for the night.

The next day we arrived at Edmonton at 10 A.M., and with the help of Mr. Lee Brintnell, of the Mackenzie Airways, I set about the matter of getting another engine. It had at first been thought that we might use one of the Mackenzie Airways' engines which was about ready after an overhaul to be installed in one of their planes, but it was finally decided that, since I could get a new engine from Pratt and Whitney's Canadian factory almost as soon as I could get the one from Mackenzie Airways, I would take the new one.

On Thursday, February 10th, I completed details in connection with the purchase of the new engine and propellers and arranged to have Mackenzie Airways fly it in to Aklavik as soon as it could be had from the factory. Although the Soviet Government had been put to great expense in connection with the search, they did not hesitate a moment in their preference for having me supplied with a new engine rather than for having me take over the second-hand machine.

During this period Sherman was, as usual, keeping his appointments, and for the first time since our tests began was failing to get even a small percentage of his recordings in line with the actual events. He had "hit" a few things, such as this, on the 8th. He said, *"See you busy at something outdoors tonight—under light from oil lamp or lanterns."*

At Fort McMurray, I was that night working with McMullen on the skis and shock absorbers of his machine. We were using oil lamps and lanterns. Sherman: *"See you talking to a short stubby little woman beneath an arched doorway."* A woman of that description had stopped me in the arched doorway of the hotel, and had asked me to sign a number of autograph books. Sherman: *"Some Geologists, or commercially interested men, have asked you to keep your eyes open for estimate of natural resources in land over which you travel—areas of mineral or other deposits."* A group of men I met at Fort McMurray were interested in the possibilities of the Canadian Arctic Islands, and, with a view to having me later go back to investigate thoroughly the mineral possibilities, had asked me to keep an eye out during the spring—if we continued our search until then, for any promising areas. But, in general, Sherman was under the impression that I was still at Aklavik, and wrote of things which might have been happening there. He did record on the 10th, *"Impression—as though unexpected development—mechanical mishap this time—would compel still another delay."* This was not so good, considering that we had discovered the mechanical trouble on the sixth, that by the tenth the men had removed the engine from the plane, and that I had arrived at Edmonton to pick up the new one.

In fairness to Sherman and our experiment, I must record that by this time Sherman must have been getting very weary and disappointed at my inability to keep him informed as to the accuracy of his results. Because of the difficulty with shortwave wireless communication that season I had no opportunity of letting Sherman know how successful he had been. He was no doubt getting very anxious to know just what the results were; and I was then, and still am amazed that he so faithfully kept to his plan of recording three times a week. On February 14th, he at last received my report on his efforts during the period between November 15th and January 27th, and it will have been seen by these recordings that he had made some very remarkable "hits."

On the 14th, Sherman was back again on the target. He recorded: *"Impression you talked three times before different interested groups since arrival at Edmonton—first time before some luncheon club—like Rotary Club—you have found a motor—you plan to take off with it tomorrow or Wednesday, if weather permits. You have dinner with three men and their wives . . . one of Edmonton's wealthiest and most prominent men has entertained you and given you some assistance relative to the expedition—word Mackenzie flashes to my mind—is there a company of that name supplying you with plane? Seem to see you as guest of Church Brotherhood . . . Sunday occasion—you called on to speak—you have appointment with two men who will take you to some plant or place where you are to see the packing of the equipment."* All of these things took place between February 10th and February 14th.

# A Fire in Point Barrow

On the 10th I spoke before the Edmonton Rotary Club and the Alberta Dairymen's Convention. I had found the new motor, and had arranged to leave on Tuesday if possible—and if not on Tuesday, then on Wednesday for certain. The express train was late, and the departure actually took place on Wednesday. On Saturday, I had dinner at Captain Stevens' house with Captain and Mrs. Stevens and two other couples—they later went to a Cavalry Ball, and I went back to the hotel. On Sunday, the 13th, at the invitation of the Reverend Dr. McDonald, pastor of the Knox Church, I spoke from his pulpit after the regular service to the Men's Brotherhood of the Church, and on Monday, at lunch, I was the guest of Mr. Pike, of the Bank of Montreal, with his two friends, Julian Garrett, president of the Northwest Utilities, and Lee Brintnell of the Mackenzie Airways.

The Mackenzie Airways was furnishing the plane which would fly the new engine back to Aklavik. I was to go with Mr. Brintnell and his workshop manager to see the engine and parts packed ready for shipment in the plane.

To every one of those incidents, which had happened and which Sherman had recorded, though not significant in themselves, I had given some intense thought. They were not of such character that Sherman's imagination could have conjured them, even if he had allowed his imagination to have play.

On the 15th, I went with Mr. Brintnell, and watched the new engine being uncrated and prepared for stowage in the plane. I found that I had not asked to have some new washers with which to connect the cylinder-head temperature gauges sent with the spare parts, and it was with some difficulty that Mr. Brintnell was able to let me have some from his stock. Sherman that night recorded: *"Intense activity preparing to leave Edmonton for Aklavik—early Wednesday morning flight in mind—you purchase a handful of what appears to be metal washers or rings . . . see color yellow or copperish with yellow tinge—wonder if this color is part of engine—odd impression at any rate."* The engine and all the spare parts were wrapped in yellow waterproofed material—all of which was exposed when the machinery was uncrated. I would scarcely have thought intensely to Sherman about that, but the yellow coloring must have attracted my close attention, and the impression of yellow was strongly in my mind.

On the 16th we left Edmonton early, and flew into Fort Simpson where we remained for the night.

# 5

## High Emotional Moments

On Thursday, we flew to Aklavik, running into some bad weather toward the end of the flight, but managing, by going around some small storms, to get in safely. The men immediately unloaded the engine, and started to assemble it. I had brought in a new propeller with the new engine, and there was some discussion among the engineers as to its suitability. It was not exactly the same pitch or length as the other propeller, and it would require careful adjustments to synchronize the motors in flight. I had considerable discussion with the engineers in respect to this.

Sherman, who has no technical knowledge in regard to planes and propellers, would hardly have *guessed* about the pitch of the new propeller, but in his report for the night of the 17th he has recorded: *"Flight back to Aklavik made successfully without incident—new propeller to go with new motor—seem to see it—something about pitch of new propeller . . . seems as though pitch of both propellers must be the same . . . I have no technical knowledge and do not know what this means."*

This is a case, as conspicuous as any which occurred during our test, in which Sherman appears to have responded directly to my thoughts—and quite apart from any response that he may have made to his own power to visualize things at a distance. He saw the propeller in his mind's eye, and he might have recorded that fact alone—which in itself would have been remarkable—but he could not have seen the difference in the pitch of the propeller, because it was so slight that it could not have been noticed by the keenest eyesight. The difference in the pitch of the propellers could not be proved except by means of a delicate instrument or by a comparison of the fine markings on each, which were concealed beneath the hub. So, to have known of my concern and discus-

# High Emotional Moments

sion with the engineers about the pitch of the propellers, Sherman must have responded to the stimuli of either my thought or of our expressed words.

From the 18th to the 22nd, the men worked as best they could on the installation of the motor. A cold spell of weather, accompanied by wind, hindered them considerably, and Dyne and Cheeseman soon had sore, stiff fingers. This was due to the fact that they would often touch a part of the machinery with their bare hands, and at each touch the frozen metal would burn and blister the skin. Some of the installation required fine adjustment which could not be done with gloved or mittened fingers. In an effort to make things more comfortable for them, we rigged up a wood stove in the tent covering the motor, but this had a disadvantage, too, since a good deal of hoar frost formed, and the warm air coming from the tent hung like a pall of smoke above the machine in calm moments, and with the wind drifted away in a wreathing column. The installation of the motor under such conditions was extremely difficult, but Dyne and Cheeseman did a splendid job in a remarkably short time. The new engine had been brought in pieces, and we found it possible, both for convenience and economy, to retain some parts of the old engine on the plane.

By Tuesday night, the 22nd, the engine was completely installed, but we decided to wait until morning before warming it up, and giving it a test.

On the night of the 21st, which was one of our contact nights, I was able to sit and review the events of the day. Sherman recorded: *"Installing of engine completed, and testing of it carried out today.* (The engine had been mounted on the framework of the plane, and some of the parts had been tested although the engine had not been "run-up.") *Very difficult job—feel that the weather delayed your work one day—someone has skinned hand and finger . . . see great clouds of smoke or vapour about the plane . . . use made of part of damaged engine. You brought back to Aklavik several boxes of cigars—some member of crew requested you to get some article—can't make out what it is—someone has toothache or sore condition of the mouth."*

Sherman had not only listed much of what I had been thinking—of the engine and the general conditions—but he had also added several things which were conspicuous in my mind, but which I had no deliberate intention of transferring to him. For instance, I had had a tooth filled while at Edmonton, and the tooth was still giving me some trouble. My mouth was sore, and the tooth "jumped" every time I trod heavily or received a jar—in fact, it had been troubling me all day. I had brought from Edmonton a special box of cigars for Wilson, one for myself, and one for the members of the crew in general.

The matter apparently referred to by Sherman in his record as *"some member of the crew requested you to get some article,"* was a radio which one of the men had asked me to order through Iversen for him. I was much worried about having Iversen place this order, because he had been successful in interesting the radio manufacturers in our effort so far as to have them furnish our radio equipment at far below manufacturing costs. This generosity on the part of the manufacturers was much appreciated, and had been accepted with thanks in connection with my scientific work and with this errand of mercy, but I did not feel justified in taking advantage of it for my own or for others' purely private benefit. I had, however, while in Edmonton, asked Iversen to make enquiries. But after I returned to Aklavik, I decided not to have him complete the order, and I was very anxious to get a message to him to that effect.

Sherman not only recorded, *"Some member of the crew wants you to get some article,"* but he included, *"You would like to get some word through to Iversen . . . wonder if this thought is in your mind tonight as you think of me?"* Sherman had also recorded during his sitting on February 14th, the night I had wired Iversen about the radio, *"You delegate . . . Iversen to secure some pieces of equipment and rush through in relation to repair job on plane's motor—also in connection with radio . . ."* It seems, beyond doubt, that Sherman in this instance, as well as in many others, was reacting to thoughts expressed and actions carried out, rather than merely recording the promptings of his imagination.

On the 22nd, Sherman recorded, *"Something about gasoline supply . . . some cans of oil seem to have been poured into plane . . . hear hum of motors, see you watching and listening intently, checking with Dyne who is up alongside new motor—he appears to have some adjustment to make—to the old motor—this getting them synchronized no easy task—hear him say finally, 'Sounds okay to me.' Last few days have been tough work—someone breaks out a bottle after work tonight, and everyone has a drink."*

All of these impressions had to do with things which had happened that day. On the 20th, I was notified that Mr. Beliakov, the Soviet Government representative at Fairbanks, had arranged to have some gasoline flown in from a supply laid down at a place named Old Crow—a police post several hundred miles away. We had expected the first load to arrive on the 22nd. It did not arrive because of some delay with the supply plane, and I was giving the matter some intense thought. It was on that day that the new engine was completely ready for a long-run test, and it was filled with oil and started up. We found, after running it slowly for some time and then at full throttle, that its maximum speed was one hundred and fifty revolutions slower than the other. This meant that some adjustment would have to be made, not only to the new motor but to the old one as well.

Dyne made these adjustments, and finally, late in the evening, we were satisfied with the operation of the engines. I celebrated the fine job Dyne and Cheeseman had done by breaking out a "bottle," which I had brought with me from Edmonton.

Sherman mentioned many other things that night which may or may not be responsive recording—such things as *"Canadian coin—King George's face dated 1903—coin in your hand—you eat steak tonight with relish, and so forth."* I did not jot down such things in my diary, but they might well have been true. Sherman *did not* record one thing that day that was emotional, or, which, at

# High Emotional Moments

times, occupied my whole attention. I find that I have noted in my diary for the 22nd, "Walked over both runways—more than twelve miles . . . blistered both feet." The boots I had been wearing had become wet from going in and out of the shed where the oil was being warmed. While out walking, they froze solid, and my feet got frostbitten and chafed before I got home. They were not badly injured, but enough to give me considerable pain. One would expect that if Sherman had "got" my tooth tenderness of a few days before, he would have felt something of my blistered feet.

After writing the foregoing, it occurs to me that after the twelve-mile walk and the work of testing the engine I did, most likely, "eat with relish" my steak, or whatever it was that I had for dinner that evening.

Sherman also recorded: *"Morgan's name flashes—you've communicated direct with him tonight."* That was a fact. Wilson usually contacted Morgan and keyed the message, but that night I *personally* keyed messages to Morgan, and received his messages to me.

It was on the night of the 24th of February that we were able to carry out with precision the test with the E.S.P. cards. There were so many difficulties, both mental and practical, in connection with this that I had little faith in getting more than the average results. At this time, and when checking the results (which we could not do immediately after the experiment, because of the difficulty with short-wave wireless communication), I find that while I shuffled the cards and turned over the first ten, looking at each one for a minute before turning the next, then turning over the same ten cards during the second ten minutes, and again turning them over in the same order during the third ten minutes, that Sherman recorded a different set of ten for each ten minutes. Of the first ten he had one "hit," number 5, which was a wave. In the second series he had three hits—numbers 1, 4, and 6: wave, cross, and circle. In the third series he had also three hits, numbers 5, 8, and 9: wave, circle, and star; but even three hits out of ten on two successive counts was not very extraordinary.

Sherman records for that night (in addition to the E.S.P. card tests), *"Wilkins now seems to have recorded the ten cards or symbols chosen in another book."* I had recorded the symbols in my diary. In fact, I had my diary before me during the test, and as I picked up the first card from the pack, I tried to think of it to the exclusion of all other thoughts, and to write its designation down in my diary as number 1, a wave, number 2, a cross, and so forth. The next time, as I turned the same cards in the same rotation, I not only wrote the words such as wave, cross, and so forth, but I drew above the words little symbols for waves, drew a cross, a circle, a square, and so on, hoping that these drawings and notations might help me concentrate.

But before me I could see all the cards, even though nine of the ten were face down. I could not consciously forget the card just turned down, as I gazed at the card just picked up. There was not time enough to disassociate the previous card, or any of the cards, from my mind. By the time I had turned them over twice I had the rotation "by heart," so that my thought of them was a jumble until I physically took one up and stared at it intently. Even then, my mind was actively conscious of the others, and was presenting them in rotation in my mind. How would it be possible for Sherman to follow those intricate jumpings of my thoughts? Apparently he did not, or at least not very successfully.

Our machine was ready again for flight, but we had decided not to test it until we had some really fine weather. On the night of the 25th, I again tried the E.S.P. cards with Sherman, and this time we agreed that I would select the first ten cards, one by one, and try to think of each one for three minutes before turning to the next. I did this; and to help me concentrate, I drew in my diary a large square full of little squares as I looked at the square on the card, and a large circle full of little circles as I looked at the circle on the card, and so forth.

Well, I find now, upon comparing my list with Sherman's, that he had tried three methods of determining the cards I had in mind. By his first method, he had two "hits," numbers 3 and 6: cross and wave. By the second method, he had three "hits," numbers 1, 7, and 8: cross, circle, and square, and by the third method, he had got two hits, number 1 and 8. Furthermore, by the second and third methods, he had got two of the cards in the right order. I do not know whether this is better than the average law of chance, or better than the average results as analyzed by Dr. Rhine in his experiments.

Even with the three-minute "hold" to one symbol, I found it difficult to concentrate on that symbol to the exclusion of all other thoughts. There could be nothing emotional about looking at a symbol in black on a white card. Just by looking at the card I could not completely hold my attention to it, when I had to keep in mind that in exactly three minutes I would have to turn over another card—one of twenty-five. I knew that in the pack were five cards with squares on them, five with circles, and so on. And there before me was my watch, and that had figures for every five minutes of the hour, and so on. The thoughts rushed by without limit. Which of the thoughts should Sherman select to record—if he was getting any impression at all from them?

On the 26th, we decided to carry out a flight over the Alaskan mountains (a thought I had had in mind for several days), in order to give the new engine a good run-in before venturing out over the ice with it. The machine was loaded with a fairly light load of gas, because I planned to stop at Old Crow, fill our tanks from the supply there, and by this means save our reserve at Aklavik. I expected also to land at Old Crow on the way back, and to bring a supply of gasoline to Aklavik. The machine was ready, but the weather was not good on Sunday the 27th, and we had to content ourselves as best we could during the delay.

On Monday, the weather was still bad over the mountains, and we grew more and more impatient. There was nothing to do however, but wait. On that night I again tried experiments with the E.S.P. cards. This time I was to go through the whole pack of cards, turning each one, and thinking of it for a minute. That night, as I tried to contact Sherman, I drew the card, turned it over, wrote its description—number 1, and so on, and drew the symbol of the card: a cross, and so on in my diary.

I now find that Sherman got seven hits out of the twenty-five shots. But it is interesting to note that in addition he recorded four other cards in which his sequence was the same as mine, although one card *later* than the order in which I had "sent" them. This may or may not be accounted for by a time lag, or by the confusion in my thought. I was not impressed with this style of experiment, and was without doubt more engaged, mentally, in thinking over the proposed flight over the mountains in search of the Russian flyers. I had hoped, rather forlornly, for good weather on the next day, Tuesday, but the weather forecasts indicated that we would not be able to take off until Wednesday.

I was looking forward to the flight, because there recently had been some reports from the Eskimos that they had heard on the Arctic coast the hum of engines on the day that Levanevsky might have reached the coast. It was just possible that he might have crashed in the Endicott Range, and I wanted to make a thorough search of that area. I had crossed the Endicott Range many times on my way from Fairbanks to Point Barrow during my earlier expeditions, but to fly back and forth along the range, over its yawning and cavernous valleys—where for miles there was no possible safe landing ground—would be interesting, although not very comfortable. We would have to fly low to see into every crook and turn, and, we would no doubt get bumped about severely by the turbulent atmosphere. Sherman, on the night of the 28th, recorded: *"Your mind has been on preparation for an imminent take-off tomorrow, if weather is favorable—for search flight over mountains—seem to see plane activity for Monday or Tuesday of this week, though know it to be in advance of proper moonlight or ordinary favorable conditions. Weather has been no good for flying, but you hoping better luck for Wednesday—feeling definite this day for action—seem to see flight taking place on this day over mountains—you can call it premonitory—but seem to be in plane experiencing bumpy air currents—looking down into deep pockets or valleys filled with snow and ice—dangerous flight—some of the mountainous areas treacherous—bad in event of forced landing—you are anxious to finish this search job—determine make up for lost time by getting full schedule in March. Get impression flights over mountains intended despite certain risk—a warm-up for much more extensive Arctic ice flights to follow later in March."*

All such things were much more definite and importantly fixed in my mind than the E.S.P. cards with which I had carried out the experiments. I am not surprised to find that Sherman got a greater percentage of "hits" in connection with expedition matters than he did with the cards, although his card recording was, I was told later, above the average.

On March 1st, we were up early—just in case the weather should look promising—but we found that the meteorological forecast was correct. There was a high wind blowing and a heavy snow-drift over the mountains. We might have taken off from Aklavik, but we would not have been able to observe clearly through the low drifting snow. Therefore, I was able to keep my schedule once more with Sherman.

We did not try any E.S.P. tests, but Sherman was able to record practically all of my conspicuous thoughts, and to describe the conditions I had experienced. He recorded— *"Real action scheduled for tomorrow—A 400 to 600 mile flight in the offing over mountains* (I had planned to go 600 miles out and 600 miles back) *investigating rumor reported by Eskimos—I feel that this rumor basis for your search this locality—have wondered consciously why you would make a flight in this region—this seems to come to me as an answer for your covering mountain territory—If Russian plane sighted you mark location and make trek by land to reach scene—seem to see you close to take-off today—see busy group around plane—cloudy weather over mountains holding up flight—you can't go far—you'd have to turn back—but tomorrow Wednesday looms as action day—weather clearing—there seems to be a mountain peak or elevation you intend to circle—in vicinity you believe Russians might be. Liquor—in that connection seem to sense commercial interest—like some firm wanting endorsement—you considering if they will offer you enough money. Feel necessity for exercise of greater caution—hope nothing forces you down on flights—suggest you check oil and gas lines leading to engine as possible source of trouble—something appears to get clogged or choked— outside atmosphere or temperature seems to have something to do with it."*

Sherman had "hit" all the conspicuous happenings and intense thoughts of the day—even to one item which I did not consciously review as I sat thinking to him. I had received a telegram from the firm of Hiram Walker asking me to lend my name to the advertising of a certain blend of whisky, and while I would have been glad to earn some extra money, the amount offered was not sufficient to tempt me. I decided that, if the amount was not raised, I would not accept the offer. This was one day of the year that I had had such an offer from a liquor firm, and it proves conclusively, I think, that there was more to Sherman's recordings than that for which his imaginings or clairvoyance, or fore-thinking could possibly account.

Another interesting record is that of *"Checking of oil and gas feed lines— appear to get clogged."* We had found that on the new engine there was not the same smoothness of operation of the automatic constant-speed device governing the propeller as there had been on the original motor. I had given it some thought, but Dyne, the engineer, seemed to think that we would have no trouble with it after it had been worked in a little. Sherman might have picked up my thoughts about that, although they were not consciously intense. Or, perhaps, Sherman might have had *some forethoughts* about the happening of the next day—something in the nature of prevision. Because certainly enough, the *next day,* during flight, the feed on the automatic control *did* clog, probably due to the low temperatures and to the fact that the propeller unit was not working satisfactorily. I have no real knowledge of the powers of clairvoyance (or prevision), and I am more inclined to believe that what Sherman recorded was in response to my discussion of the subject with Dyne prior to the time that Sherman made his record.

The planning of several flights, mentioned by Sherman, was quite correct. And the intention to circle the peak or elevation was also correct, since the Eskimos with whom I talked, and who had reported the hearing of engines on the day that Levanevsky was lost, said it seemed to them that the noise was moving in the direction of a conspicuous peak south of Barter Island. I, therefore, had intended to search that area carefully.

On Wednesday, March 2nd, we took off in flight at 9 A.M. Because of the delay and of the brightness of the weather, I decided to put more gasoline in our machine before starting, and not to land first at Old Crow. I would go at once to carry out the search, and would land at Old Crow on the way back. We flew for twelve hundred miles over the Alaskan mountains in perfect sunshine and visibility, but we failed to see any sign of Levanevsky or his plane. We flew west until we were far beyond the line of flight that Levanevsky was to take, and considerably beyond the point indicated by the Eskimos. We turned back toward Aklavik, and then again to the west over a different area, veering back again as darkness came on.

Kenyon and Cheeseman were in the cockpit, since I had decided that day to ride in the rear compartment—which afforded greater convenience both for observation and photography. Toward evening, Kenyon reported that the new engine, probably because of some trouble with the constant-speed control, was not working satisfactorily, and Cheeseman added that the *oil temperature* indicated some *clogging* of the line. As I have said, I wished to land at Old Crow that night, but because of the better facilities at Aklavik for inspecting and servicing the engine, I decided to return to our base. We reached Aklavik just after dark.

We were disappointed at not having found any trace of the missing men, but we had added many hundreds of square miles to the area searched. There was still a great deal of the mountain area we had not covered, and I hoped to be able to set out on another flight the following day.

On March 3rd, we took off at 8 A.M., and reached Old Crow at 9:15. We were to fill our tanks from the gasoline supply, but we were not quite sure just where the gasoline had been cached. So we landed in front of the village, which lies quite near the bank of the River Crow, a branch of the Yukon. The gas proved to be stored about a half-mile up the river, so we taxied the plane toward it.

As we neared the supply, the plane ran over a river bar which was only thinly covered with ice and snow. The skis and tail skid tore through the snow, and the ski-bottoms and the tail-skid surfaces were badly scored by the hard, icebound rocks beneath. At first it looked like a disaster, but we found that the damage was not great. With little difficulty we jacked up the machine, straightened out the ski-bottoms, and, after packing snow beneath them and over the rocks, we got the machine down to the ice on the river.

There the tanks were filled, and we took off on our second search flight over the Alaskan mountains. That day, we flew a distance of more than eighteen hundred miles, hawking back and forth along the Endicott Range, peeking into every draw and valley, looking for the wreckage of the Russian plane, or for evidence of men who might be stranded there. But we saw no trace of plane or men. The weather had been clear on both days of the search. We had flown, in all, three thousand miles over the mountains, and we felt sure that, if the plane had crashed there, we would have seen some trace of it. We continued our search until darkness set in, and then set out to find our way back to Old Crow. Following the river, which stood out like a broad white band between the dusky willow-covered banks, we came to Old Crow, and landed with the aid of flares which had been set out by Corporal Kirk of the Royal Northwest Mounted Police. The Corporal had been of great assistance to us, both in reporting to us the weather conditions with the aid of his short-wave radio, and in making ready our gasoline supply. Together with others at the Post, he entertained us royally that night, and we slept comfortably in the police headquarters.

Sherman, on the night of the third, recorded: *"Several times I had strange sensation of feeling myself in the air with you—passing quite low over shaggy white and shadowy peaks—below me gleaming stretches and large dark patches where the sun's rays are cut off—at other times the shadow of the plane passing over the snow—crew pointing out several herds of wild animals—several flocks of birds taking flight—you have map on knees or chart of some kind which you checked and made markings upon—you out to dinner again tonight—telling of a few of the observations made several places where prospectors have cabins."*

All of this, of course, could have been imagined by anyone familiar with the country, and knowing that we had been flying over the Endicott Range, but Sherman did not know that we had been flying over the mountains that day, and he had never seen or heard of the conditions to be found there.

He describes exactly the conditions and much of what was in my thoughts—the sights I saw, and even the action of recording, on a chart held on my knees, the flight courses which I had to keep carefully in case we might see the plane, and have to mark its locality in order that we might return to it later on foot. He spoke of the herds of caribou, and of the cabins of the prospectors—in fact we circled steeply in a tight spin, almost down to the level of one cabin in a narrow valley, just to see if there were any men in it alive, but there was no movement about the cabin, and we could not see any tracks in the snow. The cabin was probably used only during the summer months.

There was one item in Sherman's report that might have been added by Sherman's imagination: *"Several flocks of birds taking flight because of your plane's invasion of their territory,"* recorded Sherman. The only flock of birds we might

# High Emotional Moments

have disturbed were ptarmigan, a species which keeps low to the ground, especially in winter. I do not remember having seen any of them during our flight, although they might have been there. The fact which interests me is that, while Sherman's imagination might have had a little play in respect to that item, the item itself goes to prove that Sherman had no studied knowledge of the other conditions he described in the Arctic. He has never been there, and has probably read very little about it. If he had studied the conditions so far as to be able to write colorfully about it, he would have known that we would not have disturbed several large flocks of birds.

From my personal knowledge of Sherman, I know of course, that he would do no such thing as study conditions to help him at his work in receiving my thoughts, because I know that he is downright sincere and is genuinely scientific in his approach toward, and in his study of the subject of thought transference. I have referred to the bird incident, mainly as proof, if necessary, of Sherman's integrity. I do find, however, that I made a note that day of having seen the first snowbird of the season at Old Crow—observed while we were loading the gasoline into the tanks. I also talked of birds to one of the police officers at Old Crow that evening. These two facts might have suggested to Sherman the bird-recording.

On Friday, March 4th, we flew back to Aklavik. It would not be long before we could expect to take advantage of the next moonlight period, and we set about restowing our supplies in the plane, and grooming it for the next long flight over the ice.

On the 6th, we were up at 3:30 A.M. In anticipation of a favorable meteorological forecast, we got the machine ready for a take-off. Conditions at Aklavik were not good. A low fog hung over the river, and visibility there was less than a half-mile. However, if the forecast indicated good weather far out over the Arctic, we were prepared to take off, in spite of the danger such a take-off involved, or of the difficulty we might have in finding our way back to the landing field.

It was really a relief to learn that the meteorological report from Fairbanks had forecast a low ceiling and much fog for the center of the Arctic. The fog hung about Aklavik all day, and the Canadian Airways plane which had brought in the mail was unable to leave. The next morning we were up again at 3:30, and again prepared the plane; but the same thick fog-condition prevailed until late into the forenoon. It was most disappointing, and I considered leaving after the fog cleared. But since the weather forecast was bad for all areas north of 80 degrees latitude, and, as we were expected to fly to 87 degrees north—some four hundred miles into the bad weather—I restrained my impatience, and continued to hope for better weather the next day. But again we were disappointed. The weather forecasts indicated conditions over the Arctic which would have prevented us from seeing anything of the surface, and so we were once more grounded, champing at the delay.

In the meantime, during his sittings on March 7th and 8th Sherman had scored some "hits," and had also recorded a number of things which indicated that, while he was not getting all of my thoughts, he was receiving impressions of things with which I was thoughtfully concerned. During the delays owing to bad weather, I would work out the navigation problems for each succeeding day, and for the various times of the day. Depending upon the time of our start, there would be different calculations and a considerable difference in the formula required to aid me during the flight.

I planned my flights, as far as possible, in order to have the sun or the moon at an angle easy to observe when we neared the area in which we thought we might find the missing men. For the first flight in March, I had hoped to be able to be as far north as latitude 87 or 88—that is, within one hundred and fifty miles of the North Pole—and from there to zigzag east and west over a considerable area before turning back to our base over a route which would cover country other than that which we had passed on our way out. This would mean a flight of about twenty-six hundred miles over the Arctic Ocean, and the navigation for our route would need to be fairly accurate, if we were to find our base without trouble.

I had said nothing of our possible course to anyone, because it would have been useless to have announced any course before I had received the weather report for the day. But Sherman seemed to get the impression of my hopes, for he recorded: *"Your next flight to be your longest to date—around 1300 miles out and 1300 miles back—feel you intend to approach to within 150 or 200 miles of the pole."*

Sherman continued: *"Mail seems to have arrived—was tail of plane slightly damaged in bumpy landing during Alaskan mountain flights? Seem to see some work having been done in the rear of the plane."* Both these items were correct. Mail had arrived by a Canadian Airways plane, and the mechanics were busy repairing the damage done to the tail skid of the plane at Old Crow.

Sherman, also, on the night of the 7th, recorded: *"A flash of red—haven't had this in a long time."* We had, as might be recalled, an agreement that I would flash a red signal to Sherman in case of accident or illness to anyone of the crew. None of the men was ill at that time, and I don't remember flashing a red signal to Sherman. But what might have accounted for his "red" impression was the fact that at that time (a subject Sherman did not record) a freighting plane was flying cans of gasoline from Old Crow to our base. The cans were painted a *bright red,* and made a very conspicuous red-colored pile beside our silver-colored plane. I could not help thinking intensely of this red mass each time I passed our machine.

Sherman also reported on that day, *"Endorsement of some advertised product—seem to see liquor bottle which has been requested of you. Believe you acting upon offer—make decision this day."* It was on the 1st of March that I had received a message asking me if I would, for a certain sum, endorse a special blend of whisky. I replied that the price offered was not high enough. On the

# High Emotional Moments

7th I received word that the price had been raised, but while I was still in need of money for my new submarine, the amount offered did not seem to me to be high enough, and I finally decided that I would not lend my name to the advertisement.

I had given the matter some intense thought, and had almost accepted the offer. It was something that Sherman had no means of knowing anything about except through my thoughts, and it was certainly not a matter about which he would have been inclined to speculate if he were trying to imagine things to record.

On the 8th, Sherman wrote: *"Fleeting vision of your face—quite strained, intent impression as though concentrating on flight activity—something did not go as planned—snow—or sleet-like weather in some parts—see you beside plane looking up and around—there appears to be a ridge or slope beyond—of snow and ice—in ribbon-like outlines."*

In my diary I have noted for that day, "Worried about not starting out this morning—fine weather here (at Aklavik), but low clouds over Old Crow." After receiving the weather forecast which indicated, as I have noted in my diary, that "any flight over the Arctic that day would have been useless," I stood beside the two planes—the freighting-plane of the Mackenzie Airways and ours, studying the weather in the direction of Old Crow. Snow storms in quick succession were driving rapidly over the mountains which run parallel to the Mackenzie River valley. Along the mountain range, the newly driven snow spread over the stratified geological formation in such a manner as to give the whole landscape a ribbon-like appearance. I had often noticed this condition, and I had probably commented "mentally" upon it many times. But on this day I gave it particular attention, because the snow storms in that vicinity were preventing our flight. For several hours, the freighting-plane pilot and I watched the weather in that direction. He was anxious to get off on another flight, and although he was ready to start at 7 A.M., it was not until after 11 A.M. that he judged the weather over the ribbon-like mountains to be satisfactory.

Sherman reports for that day: *"Radio activity,"* and I find in my diary an entry to the effect that radio conditions were fairly good, and that I had sent long messages to Stefansson, Oumansky, and Beliakov. This might seem to be something that Sherman might have guessed, but it should be remembered that because of the general interference with radio communications that season, the sending of long radio messages was not an every-day occurrence. Sherman's recording of radio activity that day might just as well have been in response to thought transference as to a mere coincidence.

On the 9th, the weather seemed to be clearing. We looked forward to a flight on the 10th. In the early morning, before it was light enough to see the lower part of the sky clearly, the weather at Aklavik looked promising. I routed out the crew so that they could prepare the plane. The forecast of conditions over the Arctic up to latitude 80 was fair, but north of that visibility would be obscured by haze and snow drift. I decided, however, to start out, and we were in the air by 7 A.M.

Soon after we had left the Mackenzie River delta behind us, we came to a long lane of open water which stretched from Baillie Island almost to Herschel Island. The ice was much broken there, and even farther north. Many patches of open water could be seen off the southwest side of Banks Island. Curious, crescent-shaped leads of water were observed, one behind the other, opposite McClure Strait—which separates Banks Island and Prince Patrick Island—lying parallel to the islands' coast. Some distance off the shore of Prince Patrick Island was a band of rough and highly piled pack-ice, on the seaward side of which there was much open water. As far north as we went in good visibility, we saw open leads; and between latitude 79 and 80 the leads were wide and long. We had been following in the wake of a storm which was moving northeastward. At about 80 degrees north—just where our meteorological forecasters said we would—we ran into poor visibility, and soon came to thick clouds and snow. We had overtaken the storm, and in an effort to avoid it we turned to the eastward. But as far as we could see to the northward, the weather was bad.

We came down to an altitude of two hundred feet, but since even from there it was not possible to see the surface clearly or for any distance on either side of the plane, I reluctantly set our course for home.

Our two radio engineers, Wilson at Aklavik and Cooke at Barrow, had been busy working with Kenyon on the plane, and we received several "fixes" of our position from them which closely agreed with my estimate of our position. This was rather comforting, for it seemed as though we might be able to reap some advantage from our radio direction-indicating instruments. Two conditions which are likely to swing radio beams and interfere with the accuracy of radio direction-indicators are "night-effect" and the proximity of mountains. On that date, however, we were operating during the middle of the day, and far from mountains.

From our meteorological forecast, and from our observations of conditions, I judged that at a high altitude we would find a tail wind on our way home. By going up to twelve, and sometimes to fifteen thousand feet, we sped homeward at an average speed of two hundred and ten miles an hour. We landed safely after a flight which had lasted eleven hours. We had had good vision over the ice close up to 80 degrees north, and, I think, would have seen a plane or a camp had such a thing been within five miles on either side of our line of flight. We did not see any sign of the men or the plane. We were under the impression that they would be north of latitude 80, however, but no one knew just where they might be. It was possible that they might have drifted or flown south from where they were last heard from—a position within two hundred miles of the Pole.

On the morning of the 11th, the weather forecasts indicated a condition similar to the day before—good visibility up to 80 degrees north, but cloudy north of there. At 4 A.M., at Aklavik, the skies were clear. I ordered the plane warmed up, and although there was a threatening dark cloud bank to the south, I thought we might get off before it covered Aklavik. At 7 A.M., by the time the engines were warmed up and we were ready to start, the sky above us was already overcast. I hoped that we could fly under the clouds to the clear weather farther north, and although the pilot did not care for the look of things, we took off.

We had been in the air for less than a half-hour when we came to thick clouds which had settled down above the Mackenzie delta. Almost immediately ice started to gather on the wings. It was one of the few times in the Arctic that this happened, and it looked as if it would be impossible to climb up through the clouds without getting badly iced-up. It would be too dangerous to go on, but it was almost as dangerous to turn back and try to land, loaded as the machine was with 50 per cent above the normal load for a take-off. In that loaded condition the machine was not fit, even under good weather and normal conditions, to be landed. We had, of course, provided dump valves for the release of the gasoline from the extra tanks in such an emergency, but to use the valves and release gas would have been a serious matter. The gas we had, had been flown into Old Crow, and from there to our base. By the time it was in our tanks, it had cost the Soviet Government about seven dollars a gallon. I was in no mood to drop six or seven hundred gallons of the twelve hundred gallons we carried. Kenyon, the pilot, was in agreement—but, what were we to do?

It was obvious that we would have to land soon, or be forced down from the ice. In fact, it seemed that there would be no time even to use our dump valves before we would be forced down. So Kenyon bravely volunteered his opinion that he could land the machine safely with its full load in the soft snow of one of the smaller branches of the river. There the snow would be soft and deep, and we might fly the machine into it and let her settle slowly. That might be done if we could find a sufficiently long, straight stretch of river. Fortunately we saw one not far away, and with great skill and quiet composure Kenyon flew the machine at full flying speed lower and lower—beneath the level of the trees lining the banks. Then, as the skis touched the soft fluffy snow, he eased back the throttles, and the machine settled down quietly to taxiing speed. It was a fine, courageous, and skillful effort on the part of Kenyon. And, as far as I know, no other machine so overloaded has ever been landed safely with the load intact.

We were mightily pleased with the success of the effort. But as usual, if pride does not go before a fall, the fall generally comes at an unexpected moment. We had managed to turn the machine in the fairly narrow creek in which we had landed, and had then taxied back to the main branch of the river, toward our cache of supplies, and the conveniences for warming up the engines. We had almost reached the cache, when we came to a place where the river had overflowed its icy cover, and had left a band of hard, solid ice. There the tail skid caught in a husky snag, and, because of its overburden, it snapped, the tail of the machine dragging on the ice. The engines were stopped, and it looked at first as if it were a serious accident, but closer inspection showed that the damage could be repaired with the material we had on hand.

So, I set out hurriedly to get Dyne and Cheeseman to help effect repairs. The men worked all through the day and night to get the machine ready, but the weather next day and for the following two days was not good, and we were again earthbound.

I was tired, and was having my dinner that night when I should have been keeping an appointment with Sherman. What had Sherman received and recorded? For the first time since we had been conducting our experiments, he had received none of the highly emotional thoughts that I must have had on that flight of more than two thousand miles.

On the night of the 10th, he recorded: *"You seem to be about to make one of the greatest search flights of its kind ever attempted in the north—flight which will require combined light of the sun and moon"* Sherman commented, however, that he almost recorded: *"You are just completing the greatest search flight,"* and so forth, but the impression of future time seemed stronger, so he jotted down, *"You seem to be about to make."* There were several "hits," however, in Sherman's record of that night. He had written, *"Point Barrow comes strongly to my mind."* I had been much concerned with Point Barrow that day because we had been getting from there the radio messages in connection with the radio direction-indicators. Sherman continues, *"Servicing plane, busy scene."* This, too, was right, for the engineers were busy filling gas tanks and checking the engines. He had hit on another unusual incident that night, but it was not an emotional one. I had been telling a group at the dinner table about my experiences at the diamond mines of South Africa, and of how I had been given a piece of rare diamondiferous matter called "Framesite" which is a mass of small diamonds, both black and white, mixed together. When broken to show a clean surface, it presents most extraordinary beauty. I also told them that while at the mines I had been given a bag of diamond-bearing soil as a geological specimen, but that I had given my promise not to investigate or sift the soil carefully until twenty years had elapsed. The twenty years' restriction will not be up until 1942. Sherman recorded, *"Diamond mine—why I should think of this is a mystery—absurdity—but I record the impression nevertheless."* Then he wrote, *"Is there a small scar on the left side of your face near line at edge of mouth?"* I have such a scar which is not conspicuous except in cold weather. It is the result of a machine-gun bullet which struck my face during the World War, but why Sherman should comment about it that night I do not know. He probably had not noticed the scar at any time when I was face to face with him in the flesh.

The flight we had made on the 10th was reported in the *New York Times* on the 11th; and the take-off, turn back, and extraordinary landing on the 11th

# High Emotional Moments

were reported in the paper on the 12th. Sherman's next sitting after the 10th was on the 14th, so he had no opportunity of recording anything in relation to the hazardous landing and the accident to the tail skid before it was reported in the daily press. But if we look back through the records a rather startling fact may be observed.

For the night of the 8th of March, Sherman's record is prefaced with what seems a remarkable statement in the light of subsequent happenings. He stated: *"Tonight's sitting appeared to be a blend of clairvoyance and telepathy phenomena. The exact transcription of tonight's impressions are as follows: Slight break in clouds above, but dark storm clouds low on the horizon—Seems as though flight started and down at some point or turned back—plane motionless. Something did not go today as planned. Snow and sleet-like weather in some parts—seem to see it pelting the plane. I see you beside the plane looking up and around. There is activity about a tent not far distant—fleeting vision of your face, quite strained, intent impression as though concentrated upon flight activity—strange feeling in pit of stomach or solar plexus—as if I've gone through close scrape or acute experience—you concerned about something—carrying supplies."*

Those impressions, recorded by Sherman on March 8th, very accurately describes what actually took place on *March 11th:* The dark stormy clouds on the horizon—which made me hesitate and made Kenyon quite apprehensive about taking off. And, of course, we had started, turned, and come back, because of the snow and sleet-like weather and so forth.

This, and several other recordings of Sherman's, I now recall, might be regarded as premonition, or clairvoyance. I had passed them by for the reason that I am personally not particularly interested in the possibility of modern prophecy and clairvoyance. In my opinion, for many years to come, if not for always, prophecy or clairvoyance must be relegated to the speculative field—no matter what the processes governing it may be. Therefore, I prefer at present to interest myself only in the transmission of thought in relation to post-established facts. Such transmission of thought by means of some medium other than the well-known five senses may be no greater step than the step between telegraphy and radio-telegraphy. But I believe that reliable prophecy, however, is another matter.

The clairvoyant does, however, open up a field for thought in relation to time and to sequence of events. The other "clairvoyant" incidents in Sherman's record which I referred to are as follows: On December 28th, he wrote, *"January 15th comes to me as the day you actually make take-off for north regions—though you now hope to get off few days earlier in the month—This again is premonitory impression—as though thoughts jump ahead of present moment."* And on January 13th, he records, *"Am again impressed with January 15th as your probable take-off date—that conditions won't permit attempt till then, despite your present standing-by attitude."*

We would have liked to have taken off on the 13th, but the weather was bad, and our radio was giving trouble. We actually took off about midnight of the 14th, New York time and returned on the 15th.

On January 27th, Sherman wrote, *"Have impression attempted take-off around February 11th—marred by incident which will occasion further delay—again I write this as though impelled, and tuning in momentarily on future condition. Crankcase of plane comes to me suddenly—did something go wrong with it due to cold?"*

It was actually on February 6th that something *did* happen which prevented our take-off on the February 11th moonlit period. The main bearing on the engine crankcase was ground to powder—a condition resulting, no doubt, from the accident to the propeller tip on January 15th, but about which none of us *knew* until we were warming up the engine on February 6th!

While the men were at work repairing the tail skid which had been wrecked on March 11th, they discovered a crack in one of the girders of the fuselage, and had to shape some metal and rivet it to the fractured member. This was not so much of a job in itself, but in order to get at the job they had to dismantle quite a lot of the tail assembly and controls. It was a nasty, cold task, and it took them longer than expected. But by the time the weather had cleared, the plane was again in flying condition, and the tail unit stronger than it had been originally.

On the 14th, the weather over the Arctic Ocean was forecast as good up to the 80th degree. Northward from there, scattered clouds with some ground drift were predicted. It looked like a good chance to get far north, so we worked all through the night at our preparations. The plane had been left down the river where the repairs had been made, and that meant that we had to walk about three miles to reach it. But by 7:30 A.M., we had the engines warmed up, and took off with a full load of gas.

We were not sure of good visibility beyond the 80th degree, but I intended to go as far as possible that day. Finding clear weather for the first few hundred miles, we were in good spirits.

Around about 80 degrees north, we came to the scattered clouds as predicted, but they were high, and we could fly beneath them. The wind was not strong then, and we noticed that many of the leads of water we had seen on the 10th were closed or frozen over. Approaching latitude 81, we came again into fine clear weather, and we noticed a conspicuous lead running more or less in the direction I had planned to take. We followed this lead for one hundred and fifty miles—a most astonishing "river" of water in the great ice-covered sea. This lead was open to a width varying from fifty to five hundred yards, and for the whole length not an ice-bridge was seen. There were many other sections of open water near by, but I had never seen, in more than thirty thousand miles of flying over the Arctic Ocean, such a long uninterrupted stretch of open water. It was particularly interesting to me, since our custom when travelling on foot over the ice is to follow the edge of a lead when we meet it, because we

# High Emotional Moments

usually find an ice bridge within a few miles. I contemplated the difficulty we would have had that day had we been on foot, and if, when meeting that lead, we had tried to go around it.

It might be interesting to note at this writing, although I did not know it at the time of our flight, that Dr. Frederick Cook, who claimed to have travelled on foot over the ice to the North Pole, mentions in his book (and he told me about it on one of the two occasions on which I met him) that he too in that latitude—although farther to the east—followed along the edge of such a lead for more than one hundred miles. It was in the same latitude, but to the east of our course, that Dr. Cook claims to have seen and photographed islands which he named Bradley Land. There were about one hundred miles between our course and the course that Cook claims to have travelled. Our course would have been to the west of the islands, and Cook's course to the east. The lead we saw to the west might well be evidence, but it is, of course, not positive evidence, that those islands marked on the map by Dr. Cook do exist.

Near the end of the lead, close to latitude 84 north, our magnetic compass started to give trouble. I had instructed Kenyon to follow the lead carefully, following its slight twists and turns, instead of keeping a straight-line course as was our usual custom. I thought that if the Russian fliers had been forced down near that lead they might be found near its edge—a place where they might expect to get good seal hunting. We saw no trace of the missing men, but due, no doubt, to the twisting course, our compass was swinging around in circles. It seemed impossible to steady the compass. This was, in a measure, a serious matter, although I could, more or less, check our course with the general line of the snow drifts which, from our altitude, were quite conspicuous.

We had been flying at an altitude of about seven thousand feet, but at about latitude 85 we ran into some low clouds, and we were forced down and down toward the surface. Because we seemed to be getting quite near to the ground long before our altimeter indicated that we were near ground level, I decided to go on down close to the surface in order to check the instrument.

We were still, I believe, about five hundred feet about the ice when our altimeter registered about fifteen hundred feet above sea level. It was lucky for us that we could observe the surface, for if we had needed to be guided by the instruments alone—what with the compass swinging and the altimeter more than a thousand feet in error—we might well have flown right into the ice in bad visibility. It was certainly no place in which to be playing about in bad weather. But the weather ahead was still fair, although we were approaching an area of haze which more or less blanketed the surface. That, of course, meant wind at low levels, and, for all we knew, at high levels, too. But we continued on until we were almost 88 degrees north, according to my reckoning—within one hundred and fifty miles of the Pole. Then, since we had been out about nine hours, and had, we estimated, gas enough for twenty hours when we started, we thought we had better not go farther north. We zigzagged about for a short while, searching the area as best we could, then came down low over the ice to follow the snow drifts carefully in order to keep a steady course until our magnetic compass settled. Our gyro compass seemed to be in working order, and when, after a few minutes of straight flying, the magnetic compass stopped its erratic movements, we climbed and headed for home.

We had been in frequent wireless communication with Aklavik, and at times with Point Barrow. The radio engineers had been checking our position with their radio direction-indicators, and had been putting us first on one side and then on the other of my estimated position. There was, evidently, something wrong. I might well have been wrong in my calculations, but I could hardly have been so far wrong as to be first as far to the right, and then as far to the left of the course as the radio direction-indicators showed. In fact, it would have been impossible for the plane to have travelled so far in a crooked course as to be each time where the indicators put it. For those reasons, I determined to trust to my own luck and judgment, and to depend upon my own navigation. At least, I would not be swinging about to my complete confusion while trying to follow the radio directions.

At Aklavik, to which base we wished to return, the weather had been getting steadily worse throughout the day, and by the time we turned back, it was really bad. There were low clouds and fog, and visibility was less than a half-mile. It would be quite dark, except for the moon which was already past its full, by the time we reached our base—if we could reach and find it. So I contemplated a forced landing somewhere out on the ice rather than risking a messy landing in the fog in the timber-strewn areas about Aklavik. I had had experience before of landing planes on the Arctic ice, and of waiting until the weather cleared—experience, in fact, of even landing out of gas, and leaving the plane behind to walk home. But Kenyon had not had that experience, and he was a little worried at the prospect. To land and await clear weather did not much concern me, however, but to continue flying until we ran out of gas, and, perhaps, still be above the thick clouds and fog did worry me. I did not intend to be caught in that position if I could avoid it.

But soon we came to clouds which banked higher and higher. I could see from the formation of the cloud tops that they were being driven by a strong wind, but above them we could get no indication of the trend of the air in which we flew. I had to guess that. We had been in a wind which gave us a five-degree drift, when I had last been able to get a drift sight, but that gave very little indication as to what the wind force and direction might be nearer home.

The radio engineers continued their observations, but as it was getting toward the time of evening when "night effect" would interfere with the accuracy of their observations, we could not, for the time being, count on very much help from them.

In about latitude 75—some three hundred miles north of our base, we ran

# High Emotional Moments

into really bad weather. We had to climb to nearly fifteen thousand feet to get above the clouds in which ice might form on the machine. Then, as we neared the coast, I hoped that we might get down through the clouds while we were still over the sea-ice. We tried to descend, and managed, by swinging and circling between the billowy masses—which were ghost-like in the moonlight—to get through the top layer, only to find another layer beneath. The lower layer was thick and tightly packed. There seemed to be no chance of getting down through it; and to make an attempt to do so—knowing that low-fog conditions existed at Aklavik, and with our altimeter adjustment so uncertain—would have been foolhardy.

The only thing to do was to keep above that layer. It was too late to use my plan of coming down to the ice to await clear weather. We would have to go on, or turn back—we did not know how far to the north—to find clear weather. Since we did not have enough gas to enable us to turn back far, we were in a pretty pickle.

We were at an altitude of five thousand feet, between two layers of cloud. The mountains ahead, near Aklavik and to the west, we knew to be over seven thousand feet high. We knew that at that level there was cloud. The only thing to do was to go up above the upper-level cloud mass, wondering if we could get up through it without icing up. And by the time we got into clear air, after our turning and twisting, what would be our position? We had zigzagged at high speed to come down through the mass, and we would have to do the same to get up through it. But we finally climbed out into fairly clear atmosphere. There was a little ice on the machine, but not enough to give us trouble.

I gave Kenyon a course, and settled down to plot the many turns and twists we had made, trying to fix our position. Then, after about a half-hour, I suddenly noticed that Kenyon had mistaken the figures that I had given him, and was flying in a direction 90 degrees off, or at right angles to the course I wanted to follow. To have done that for a half-hour meant that we were then a hundred miles to the west of where I expected to be, and no farther south. We would also be almost a hundred miles farther away from Aklavik. Gas was none too plentiful, and still, at Aklavik, according to the last radio report, fog blanketed the ground.

I began to think that here was another time when I was facing a serious difficulty which I had no reason to believe that I could surmount.

Ahead of us was a mountainous area—and beyond that was forest-land, all covered with fog. Behind us, and to each side, lay the rough Arctic pack-ice, and lanes of open water. There seemed to be little to choose from in the matter of safety. I was reminded of the time when, with Carl Ben Eielson as pilot, I flew over the rough pack-ice to the northwestward of Point Barrow in 1927. We were on that occasion short of gas, and it seemed, because of a raging storm, that we would not be able to find our landing field at Point Barrow. All we could do was to fly against the wind and against hope, too, toward our base, for we expected that, when our gas finally gave out, it would be the end of our flight and the end of us.

On that occasion we had been saved as though by a miracle. When the gas had given out, we had glided down through the storm—through the driving snow—blind, so far as conditions about us were concerned, for we could not see outside the cabin of the plane. It was pitch dark, and the windows were covered with thick snow and ice. We braced ourselves against the empty gas tanks, and waited for what might come. Neither Eielson or I had had any real reason to believe that we would come out of it alive. It was for fate to decide—whatever came was inevitable, and we could do nothing about it except keep the machine on an even keel and wait for the crash. Whether it would be on rough ice, water, or land we did not know. We sensed the plane nearing the surface—it would be only a few moments, and then the end, perhaps. We were resigned, and, I remember, not in the least bit flurried. We had had time to calm ourselves and wait for the inevitable—and it came. Just as we were about to crash into a high ridge of ice boulders, the tip of the wing on one side of the plane had caught in a snag of ice, and the machine had swung almost at right angles, hitting the ice on the only smooth patch to be found for several hundred yards around. We had not known this, of course, until daylight had come, and the storm had temporarily subsided. But there we were, safe, unscathed, on the ice at a position I later fixed as seventy miles from our base. We could walk that distance to our base—and we did.

But in the Electra, with Kenyon, I felt it would hardly be likely that I would for the second time have such luck or divine protection.

We were probably "in for it" this time. Still, there was no use worrying, and we still had some gas left. Suddenly, through the billowing clouds, I saw the peaks of the mountain range flashing in the moonlight. The clouds were lower there, and at last we would be able to fix our position. I knew that we would have to follow those mountains to the eastward until we came to the end of them, and that then we could come down almost to sea level—even through the clouds—with safety, for we knew that to the east of the mountains was the river delta on which there were no high hills. Much of the river delta was covered with low, dark-colored trees which we would see at least a few feet before we hit them. By approaching the surface in a long glide we could contact the surface with safety—a job we could not have done over the mountains or the sea-ice. We would have been into the mountains before we knew it, and because of the fog and the flat whiteness of the ice and the errors of our altimeter we might have flown right into the ice before we could pull up.

Kenyon by this time was tired, tired almost beyond further effort. He had a splitting headache, which made it almost impossible for him to receive radio messages, so he switched off the instrument as being useless.

I, too, believed that we could better depend upon our own ability to locate our base while flying to the eastward. Clouds almost entirely shrouded the mountains, but we caught a glimpse of them here and there. And then, as we came close to their eastern end, the topmost surface of the clouds was lower, although they seemed thick and tight to the ground. However, we could now fly safely on instruments, and give Kenyon a rest for a few minutes before he had to tackle the difficult and dangerous job of gliding down through the clouds until we saw the tree tops.

I picked up the radio receiver, and heard the Aklavik report. It was "Still thick fog—visibility less than a half-mile." We had, by this time, been out for more than nineteen hours. We had been up and busy for more than thirty hours, and we had gas for only about forty-five minutes left.

It was time for us to get down, or be forced down. We turned out toward the river mouth where the trees were low, and where we might—when we contacted the ground, and by flying westward—pick up the broad, white-surfaced river where it wreathed its way among the trees.

Slowly we sank into the soft downy cloud mass, my eyes straining for a sight of the willows, and Kenyon's eyes glued to the speed indicator, the rate of climb and descent meter, and the altimeter. With the throttle back until we were only just maintaining flying speed, we glided at the minimum angle. In spite of this, and as if with a blow on the face, the black-looking willow trees zoomed out of the misty whiteness. I gave Kenyon the signal, he skillfully eased back on the wheel, and we glided over the tree tops. Now we had to fly low until we came to the river—or to the mountains, if we were already west of the river. We were not sure, although I was nearly certain that we were far enough to the eastward. Almost at once we came to a broad white mass between the trees—it *was* the river, and with a quick banking turn Kenyon swung the machine onto its course.

There are three large branches of the Mackenzie River which pour into the sea. Were we on the branch on which Aklavik was situated? We were not sure, but we were certain of one thing: that we could land if necessary, and with safety, on the river ice—even if we were over the wrong branch. It was almost like receiving a reprieve from a death sentence. While Kenyon skimmed the ice with the machine, making swift, steep banking turns to follow its windings, I struggled to check the course we were following, trying to fit it to our map of the rivers. I had hardly time to announce that we seemed to be on the right track, when the glimmer of flares which our men had set up on the ice gave us a heart-throbbing thrill.

At full throttle Kenyon speeded over the lights, went into a steep climbing turn, and glided back and down along the beacons to a smooth landing. Our friends flocked about us. We had been out nearly twenty hours. We had flown twenty-six hundred and fifty miles over the Arctic Ocean, and, all-told, a distance of more than three thousand miles. When our gas tanks were checked next morning, we found that we had gasoline left for only about thirty more miles of flight!

# 6

## Sherman's Telepathic Record

ALTHOUGH OUR LONG FLIGHT on March 14th had covered much of the area allotted for search, I was most anxious to make a flight along the shores of the northern Canadian islands—for two reasons. It was just possible that Levanevsky, if he were still alive, might with his companions have walked to one of the islands, and might be found along the coast. Also, I had been impressed with Sherman's ability to sense or "see" the things that had happened to me, and I wanted, so long as it was within my line of search, to have a look at the area between two of the islands where three people independently had "seen" Levanevsky's plane. I was not then, and am still not, so far persuaded that the phenomenon Sherman and I had been interested in—thought transference—is reliable enough to make it advisable for a person to be influenced by its recordings; and I would not, on that count alone, undertake a search in a vicinity indicated by so-called receptives. But so long as I had to keep up the search, and had already covered other likely areas without satisfactory result, I would have liked to cover the area indicated by people who claimed that they had "sensed" the position in which Levanevsky was to be found.

The men hurriedly set about preparing the plane for another long flight, and it was ready by the time we were to receive the next general weather forecast of conditions in the far north. The forecasts, however, indicated that the weather over all the Arctic Ocean was bad, and it was not expected to improve for several days. We were all very disappointed, and Cheeseman was particularly so, for he was looking forward to a share in the long-distance flying.

On the 15th of March, I received the following message:

"I am empowered by my government to express our sincere gratitude to you, to Air-Commodore Hollick-Kenyon, and to your other colleagues for your generous efforts and sacrifice of time and personal considerations in searching for the lost Soviet Flyers. I wish to inform you that my government has decided to stop all searches from the Alaskan side. Please inform me of your plans for closing the expedition." It was signed: Ambassador Troyanovsky, Soviet Embassy, Washington.

Well, that message was to bring an end to our search. But what did Sherman receive in connection with our late efforts?

Sherman, writing on the night of March 14th, while we were still in flight, and when, on the homeward journey, I was able to think of him, recorded: *"There has been plane activity today, but tomorrow, Wednesday, seems to me to mark one of your greatest flights to date.* (It would be Wednesday by the time we returned to Aklavik.) *Believe you discovered crack of framework in tail of fuselage which also needed repair.* (This too was right.) *Seem to see you manipulating hand pump of some sort in flight—one of the engines emitting black spouts of smoke—sharp detonations from the motor—uneven, choked sound—As though some carburetor trouble—gas feed."* This too was correct. Several times we had to change over from one gasoline tank to the other, and Kenyon usually did that, pumping-up the pressure with a hand pump. But once, on the way home, Kenyon was tired and busy at something just when a tank ran dry—a red signal flashing on the instrument board giving notice of this—and I was a little late in switching over to the next tank, having to pump furiously to get up pressure before the engines stopped. The engines did splutter, cough, and backfire before picking up regular speed. It was thrilling for a few moments; for some seconds I had wondered if I had switched on to the right tank, and if the engines would stop before I could get a supply of gas to them. The whole episode was over in a few minutes, but the incident had occupied my full and intense thought as much as anything had that day. It was soon forgotten, however, and I doubt that I ever consciously directed a thought of the matter to Sherman. I was thinking more about the success we had had in getting so far north, and in covering so much of the area we had been asked to search. It should be remembered that, at the time of Sherman's sitting, we had not come to the thrilling experience of coming down through the fog and landing in the glare of the firepots. Those things did not occur until three hours after Sherman had completed his notes for that night.

Sherman went on, *"Ice on plane—thin coating which you watch closely—see plane circling low over certain area—icy waste with several open stretches of water—impression 86—115 location."* IT WAS AT ABOUT THAT LATITUDE AND LONGITUDE THAT WE FLEW LOW DOWN OVER THE ICE TO TEST OUR ALTIMETER. Sherman continues, *"Strange impression of immediate, past, present, and future time impressions crowding together tonight—difficult to detect which are which. See another flight turned back because of weather conditions turning against you—and strenuous effort to reach base before snow and sleet close in on you."* I was, of course, at the time of my appointment with Sherman, much worried about the snow, ice, and fog conditions reported from Aklavik. Sherman reports, *"Spots of oil on the windshield,"* and I remember that on several occasions I was annoyed at having to wipe spots of oil from the windshield in front of me, while making that long flight over the ice.

Sherman kept his appointment on the night of the 15th. On that night I was annoyed, because of the fact that the weather forecasts indicated that we would not be able to go out over the ice again for several days. I had not then received the message from the Soviet Ambassador saying that no more flights were to be made over the ice. Sherman recorded: *"Something galling you— strong wind impression—may be fact that weather over Arctic keeping you from setting forth on an extensive flight planned—unsettled feeling—obstruction of some sort—can't feel flight today—stopped sensation—need of cobbler—repair some boots for Arctic use—see some work being done on them."* All this was correct— even to the boots—for I had taken some boots from the store for Cheeseman, who was to make the next flight with me, and the boots needed some alteration.

Sherman recorded several other things such as, *"You have given instructions to someone to pay money,"* and, *"feeling that a doctor has been called in for some purpose."* These things may have happened, but I have no record of them in my diary, and have no recollection of them now.

On the 16th, I was still hopeful that I might make the flight along the edge of the northern Canadian islands. I had made such suggestion to the Soviet Ambassador, and I busied myself with the navigation problems connected with such a flight. There was no answer from the Ambassador until the afternoon of the 17th, when I received word that the order to stop all flights over the ice must stand. All flights and flying activities in connection with the search were to be stopped on the Russian as well as the Alaskan side. So I at once set about preparations for our departure from Aklavik. I was disappointed at not being able to make the search along the Canadian islands, but since the Soviet Government was stopping all search, I was very glad to be getting ready to start for home. It would enable me to keep my promise to Lincoln Ellsworth to be back in New York by the 25th of March. I did not realize when I gave that promise—in October—that I would be able to keep it by arriving in New York on that very day.

As soon as I received definite instructions to leave Aklavik, I put all the men to work packing up. I went the rounds to settle all my accounts at the various stores, and then had a farewell dinner with my good friends, Police Inspector Curleigh and his wife. All our friends had been considerably worried about our adventure on the last flight, and they were all exceedingly happy that we would have to take no more chances over the Arctic Ocean. We could not help but join them in that thought, for I had undertaken more hazardous flying during the search than I ever had taken before, or expect to take in the future.

But so far as Sherman knew, I was still carrying on with the plans for search.

He had no means of knowing of the instructions I had received from the Soviet Government, and might have recorded, as well as anything else, that I was still busy with preparations for another flight northward. He did not get the impression from me that I was packing up to leave for the south, but in a long record he placed some interesting hits and near hits. He wrote: *"For reason I can't determine at the moment I get an exhilarating feeling from you as I sit tonight—an impression that you feel particularly good about something—somewhat excited and lively atmosphere—others of the crew seem to reflect the same reaction—feel you on last leg of flight concentration now—with several unexpected and spectacular things to happen. Little confusion in mind between flights actually made and flights planned, since your mind strongly on both."* Then Sherman seemed to get a general resume of the whole flight as we experienced it on the night of the 14th–15th.

I wonder if that was because I was, at the time he was sitting, and when I should have been keeping my appointment with him, at dinner with the Curleighs, and telling them of our experiences during the last effort? They had been extremely worried about us, and had sat up most of the night waiting for us to return. This was the first opportunity I had had to tell them all about the trip. Sherman recorded a great many of my vivid impressions: *"flying through haze—compass and direction finder entirely 'batty'—flying blind—see flares on ice—lighting up the landing field, and so forth."*

On the 18th of March, we were up early, and we finished our job of packing the equipment which was, for the time being, to be left at Aklavik. It would be shipped out by steamer in the summer.

By 7:40 A.M., we had our engines warmed up, and although there was a wind blowing at nearly fifty miles an hour over some of the route we would follow on our course to Edmonton, we started up with the intention of flying right through in one day, if possible. The weather in the south had been warming up, and since there was no snow left on the ground at the airport in Edmonton, we would have to land with our ski equipment on the rotting ice of a lake about twenty miles from the town.

Although we flew through much bad weather, and had to fly blind and on instruments much of the way, the return flight was child's play compared to the flights we had made over the uncharted Arctic Ocean.

Captain "Wop" May of the Canadian Airways, and others, met us at the lake. We soon had the machine hauled out of the mushy ice to a firm foundation, where the skis could be removed, and the wheels fitted. That night we slept comfortably in luxurious hotel surroundings, and realized that our search for Levanevsky was ended. We had seen no trace of the lost flyers, but we had—in the amount of flying we had done throughout the fall and winter, covering a distance in the air over the Arctic Ocean equal to about fifteen trans-Arctic flights—proved that trans-Arctic flying was possible without more than a normal number of accidents.

Only three trans-Arctic flights by airplane had hitherto been made—my flight with Carl Ben Eielson in 1928, and the two flights made by the Russians in 1937. But after the number of flights made by the Russians from their side in search for their missing comrades, and the flights we had made, it could be truly said that trans-Arctic flying had been successfully demonstrated.

We had put off flights on many days because of bad weather, but it is important to remember that much of such "bad weather" was bad weather from the point of view of close search over the surface of the ice, not necessarily bad weather for a trans-Arctic flight. It has been definitely proved that, when such air routes are necessary, it will be just as easy to fly over the North Pole route as to fly over any other. There will remain, for some time, of course, if not forever, some natural hazards such as may be found in flying over any other ocean. On the other hand, some of the hazards of ocean flying are not to be found over the Arctic. For instance, although we had not found it necessary to land while carrying out the search for Levanevsky, I had, on previous expeditions, landed on the Arctic Ocean because of engine trouble, and walked home—a feat which could not be carried out if I had landed on any other ocean.

With our search over, we had then only to fit the wheels to the machine, and fly it back to New York for delivery to the Soviet representatives, who would crate it and forward it to Russia.

With the two machines we had used, we had experienced little trouble because of the conditions in which we flew. Our only serious mishap had been the damage to the propeller caused by striking a stick. That had been inadvertently stuck in the snow on the runway by an Eskimo, who had come to see the take-off and had crossed the field after it had been inspected. The damage to the tail of the machine which we experienced in taxiing with an extreme overload would not have occurred to a normally loaded machine. We had beyond doubt proved that the Arctic, of itself, offered no insurmountable difficulties in the way of trans-Arctic aerial transportation.

Several of the members of the expedition were paid off at Edmonton and sent home by train, while Kenyon and I flew the machine to Winnipeg, and from there made the flight to New York in one day, stopping at Minneapolis for fuel. We reached New York on the evening of March 25th.

Because of the instructions to terminate the search and our expedition, and because we had lost no time in packing up our supplies and returning to New York, Sherman was not aware that we had left Aklavik and continued his sittings.

On the 21st, when we were at Edmonton, he recorded: *"Weather hasn't been favourable lately—clouds and haze and wind—your thoughts are on finishing up and getting away—unsettled sensation covering last few days—break up or crushing of ice around Aklavik—shifting equipment—changes made."* With his *conscious* mind still impressed with the belief that we were still at Aklavik, Sherman seemed to be getting a confused summary of many things that had

happened since his last sitting. Many of the things he recorded might be directly associated with my thoughts about the happenings, for since my return to Edmonton I had been busy writing a complete report of our experiences for the Soviet Government. I was again too busy to keep my appointment with Sherman, and his records seem to lack evidence of reaction to direct cooperation. On the 22nd, Sherman, still under the conscious impression that we were at Aklavik—although we had in reality reached Edmonton on the 18th—recorded, *"Packing up—thought seems to come to me—equipment you no longer need—arrangements being made for a supply plane to aid in carrying supplies back when expedition job completed."*

The Soviet Government had, as a matter of fact, offered to send a special supply plane to Aklavik to aid us in our return, but I had assured them that such expense was unnecessary. All of the men of the expedition could crowd into the one machine, and the supplies could be sent back by steamer in the summer.

I was, as I have said, writing my reports on the 22nd and 23rd of March, and I was naturally turning over in my mind all the events of the expedition, and particularly the events of the last few days. I was still feeling disappointed at not having made the flight along the Canadian islands. When I first read Sherman's recordings for those two days, it appeared to me as if he was reacting more than he had before to his conscious knowledge—or, that perhaps he was being influenced by his imagination. But, upon analysis, his reports really seem to be spattered with the intense thoughts I was experiencing while writing my report for the government.

Sherman recorded: *"Your crew being notified of approximate time you planning to finish up search efforts—one particular flight you want to make—wheels impression—changing back from skis to wheels—preparatory to flying plane back—turning it over to Russian Government, and so forth."* All these things were strong in my thoughts, and Sherman would hardly guess about them in preference to the many other things he might have chosen.

Then, on the 23rd, he wrote: *"Wilkins feeling exhilarated about something—this feeling reflected by entire crew. Would not surprise me to learn that you are all at Edmonton—that you have been there for several days—perhaps since last Friday.* (Friday was the day we arrived at Edmonton.) *I feel relieved as I record this—as though at last I have pulled through what you have been trying to send to me telepathically—Feeling of you being back in hotel—windy trip.* (I remember our flight through fifty-mile-an-hour wind on my way back from Aklavik.) *Wheels on plane—believe you have put them on now—have you flown to Winnipeg—feel flight action—as though you heading for New York—you have covered over 20,000 miles searching for the lost Russian flyers—feel you not quite satisfied—a search flight you would have liked to have made—break up of ice—spring impression—relief mentally and physically is enormous—feel much less tension from you—which has been terrific last few weeks—can't explain how I feel,* but sure I'm right about you not being at Aklavik—see your base there deserted—Your man at Point Barrow getting ready to leave there by plane also—you have been wined and dined—made a speech recounting your experience. You've thought of me in more relaxed way in last few days with different emotional urge—you have gone over additional recorded impressions of mine picked up at Edmonton. Would not be surprised to see you in New York as early as Saturday—I get feeling your eagerness to be about other things—new submarine, and so forth."

Each one of those items may be regarded as a direct "hit," conceived by Sherman without any foreknowledge of the actual happenings, and with very little to go on so far as the possibilities were concerned.

He did not know what we were going to do with the plane—whether I might sell it in Canada, as was once proposed, or whether I would fly it back to New York, and turn it over to the Russian Government.

I can well understand Sherman's feeling of relief—whether he sensed it from me, or whether he felt it as a personal matter. He had been ill, and, if—as is his impression as well as mine—he was in some way subject to feeling my thoughts and the conditions about me, he had been reacting not only to his own illness, but to my personal worries and sufferings, as well. It would be as if he were living the lives of two individuals, and taking on the ills of mind and body of both. I had not been physically ill, but I had been much worried in mind, both because of the approaching end to the season in which the search could be carried on, and to the probability of not being able to find the lost aviators. But once the decision to bring the search to a close had been made, I was greatly relieved.

I wanted to get back in time to organize my next expedition to the Antarctic, scheduled to take place that year.

We left Edmonton in our plane on the 24th, and flew back to Winnipeg. I had arranged to stop there overnight, in order that I might, in person, thank the many people who had been of great assistance to us in carrying out our work in the Arctic. A general meeting of our friends and helpers was arranged, and I told them something of our efforts, which, I said, could not have been carried out without the help of many people and many governments. I said that our failure to find the missing men need not be conclusive evidence that they were dead—they might even then be walking out over the ice to some Canadian island, or along the shores of those islands, and while they might be still alive at that time, they still might not reach the civilized world alive. The men of Franklin's expedition had starved to death while walking southward along the shores of the Canadian islands, because they had not the skill, nor the technical knowledge of how to live off the country. Stefansson had lived off the country, and I believe that with my experience I could, also. But I was not sure that the Russian fliers had had the actual experience necessary for them to take care of themselves on the sea-ice or on the land, and it might be months or even years before we knew their fate.

All of this Sherman recorded while sitting on the night of the 24th. His record starts: *"Feeling that you left Aklavik some days ago—first to Edmonton where you changed skis for wheels—you in Winnipeg now—Fort Garry Hotel—you make address of thanks—thanks to different countries for cooperation. Did you say in your talk you consider it possible that the Russian filers could still be alive and making way across drifting floes toward land—though strong probability Russian fliers have perished—Mayor of Winnipeg welcomes you at field—you communicate Point Barrow—is there some boat that is to pick up your supplies later—feel flight from Winnipeg imminent—perhaps take off tomorrow."*

I had, as will have been seen in the foregoing pages, done and said all of these things—even to the communication with Point Barrow, and to arranging to have the supplies brought back by boat in the summer. Furthermore, after receiving a batch of Sherman's records at Edmonton, I had resumed my habit of willing things to him as they occurred.

Whether it was as a result of that "willing" or thinking intensely that it became possible for Sherman to get a much greater percentage of hits on March 23rd and 24th than he did on the 17th, the 21st, and the 22nd, I do not know, although that certainly appears to be the case. It is definitely true that in the earlier days I had slackened at my job. Because the search was over and there was no longer the need to use our experiment in case of a forced landing on the ice, and because of the rush to return before the break-up of the ice on the lake near Edmonton, I had neglected my part of the experiment.

We may not have proved that telepathy between two people at some distance apart is beyond doubt and by arrangement possible, but I was personally pleased to have engaged in the experiment, and feel that we have proved that the subject is entirely worthy of further attention.

For Harold Sherman, because of his tremendous efforts in this connection, and because of his honest, scientific approach toward a job which must have caused him great anxiety and not a little material disadvantage, I have the greatest respect and gratitude.

Part Two

# My Story As Receiver

By Harold M. Sherman

# 1

## What Happened to Me

As with many people, the first experience which impressed upon me the possibility of mind communicating with mind was a dramatic one.

I had switched on the electric light in my room hundreds of times. There was no apparent reason why, this late afternoon in the spring of 1915, I should suddenly be seized with a feeling of apprehension as I took hold of the bulb.

"Don't turn on the light!" a voice, in my inner ear, seemed to say.

I stood for a moment, hesitant, my fingers on the switch. I had been seated at my typewriter desk, typing, and the sun had dropped below the western horizon. The giant shadow of dusk was moving over the landscape.

"All right," I decided, wondering why I was surrendering to the impulse, "I won't turn the light on just yet. I can still see."

And so I resumed my seat and my typing. But it soon became difficult to distinguish the keys. I arose once more and put my hand on the bulb.

*"Don't turn on the light!"*

The impulse was stronger this time—a positive warning! I instinctively let go the switch. I had never felt this way before—almost as though some invisible presence had spoken to me, not in an audible voice, but in words which rang out in consciousness.

It was absurd for me to let an inexplicable urge like that paralyze me. And yet I could not bring myself to go against the impulse. Instead, I pulled my typewriter desk nearer the window, and turned it about so that the dwindling light would fall upon the typewriter keys. In this manner, I was able to type for perhaps five more minutes.

By this time, the light simply *had* to be turned on. I was not going to put up with any more nonsense from this strange whim of my mind.

My hand was on the switch above the bulb, when I heard heavy footsteps come running up onto our front porch. There was no one at home but me. A fist pounded excitedly on our front door, and then the bell commenced ringing furiously.

I hurried downstairs, and flung open the door. An electric-light linesman burst in.

"Don't turn on the lights!" he gasped. "There's a high-voltage line down across your wires outside . . . and there's no telling what might . . . !"

I was told afterward that little or nothing probably would have happened. But this young linesman was taking no chances. Investigating a "trouble call," and finding what was wrong, he had raced to three residences whose lines were covered by the high-voltage wire. Our house was the last one he reached, and at each place he had shouted: "Don't turn on the lights!"

As I reflected upon my own feelings afterward, I began to wonder. Had I, in some unaccountable way, picked up the frenzied thoughts of this linesman? Had my mind, the instant it was focused upon the light, tuned in upon his "thought warning"?

The evidence seemed to permit no other deduction. What had happened could hardly be set down as mere coincidence. I was forced to conclude that, if thoughts could travel through space under certain conditions, this had been a clear-cut case of thought transference.

From that moment on, there was born in me an ever-present desire to know more about the inner powers of mind, to learn what I could through such experiments and research as I could accomplish privately with my own mind, and to add to this meager knowledge by study and contact with those who could demonstrate what was then regarded as supernormal mental abilities.

I made it my business to seek out those who were reported to possess unusual powers of mind, in order to determine, if I could, under what mental laws they must be operating which enabled them to receive impressions, more or less at will, from the consciousness of others.

My quest brought me in touch with many individuals who were practicing trickery, either for money or for self-notoriety. But, in a few outstanding instances, I came across people capable of performing real phenomena of one kind or another.

One of these was Jacques Romano.

I discovered, through him, that *all* of us are constantly being bombarded by thoughts from the minds of others, but that most of us either push aside or give these thoughts no recognition, or put them down to mere imaginings. Romano, I found, spoke aloud those fleeting impressions, and, by long cultivation, had learned to sense in a large percentage of cases whether the impressions that came to him were genuine or fancied.

"*When* these impressions come to you," I asked, "what is the mental sensation you get?"

"As though someone had told me of certain happenings, a long time ago," Romano replied, "and I am trying to recall them. If the occurrence was a big moment in the life of the person, I seem to be able to recollect it more easily. I often get an accompanying emotional reaction just like the individual experienced when the happening took place . . . and I simply put what I feel into words, in describing the incident. Of course, I see a series of pictures in my mind's eye, too . . . sometimes not in the regular order . . . all fused together . . . which I have to separate and relate, one to the other, as many people have to do when they are recalling to memory an event which has grown hazy in their minds."

Romano has stated to me on many occasions that everything that happens to us is recorded in consciousness, and that, just as a perfect memory can recall it, so the mind of one who knows how to "attune himself to the mind of another" can tap this memory stream.

Quite often, in getting an impression, Romano will be able to place the incident he is describing in the year it happened, and even the month—or actual day!

"I am not reading your mind," he will point out to the person to whom he is giving the impression, "since you have not thought of this event for years. But everything that *has* happened to you is *co-existent* in consciousness . . . and I feel these conditions and occurrences when I 'tune in,' just as though you yourself were recalling them!"

On pressing him to tell me, as nearly as he could, how he determines the time element, Romano answered:

"Investigators of mental phenomena are usually surprised at my accuracy in sensing time with relation to impressions I am receiving. How I do it is difficult to describe. Perhaps I can get it across to you in this way: when I commence to sense a condition, I feel like reaching out, mentally, into time. I start reaching back, back . . . into the past . . . until something inside me says, 'Stop!' . . . and I stop at the year I have in mind at that moment.

"If I don't feel quite satisfied inwardly, I reach back a little farther, and, generally, I get a positive flash that I have settled upon the approximate time when the event occurred . . . then I proceed to tell what happened at that time. Occasionally I will over-reach . . . and have to advance the date. But this running back or forward in time, in my mind, only takes a few seconds before I have decided upon the period in which I feel the incident took place. I will usually name the year, positively, and state about how old the person was then. What gives me this 'time sense' or is behind my being able to do it, I cannot explain."

Nor can anyone else, as yet, satisfactorily.

I found that Romano often could not give an acceptable scientific explanation for the undoubted powers of mind which he demonstrated. I was forced to make my own observations and deductions. It has been my belief for years that these mental faculties, if they exist, can be demonstrated, in greater or less degree, by any average human who will devote the time and energy necessary to developing these latent powers in himself.

To prove to myself that Romano's powers *were* genuine, I decided early in my association with him that I must endeavor to duplicate them—at least in part.

Many times, when he was not aware of it, and when he was giving impressions of people, I would be sitting in the room, trying to "tune in" on them also, checking what impressions came to me with the comments that Romano would make concerning them.

Infrequently, at first, I would be startled to hear him describing an event that my own mind had "tuned in on," some outstanding happening in the life of the person which had been strongly recorded in consciousness. I would seldom make these impressions known, since I was experimenting for my own inner satisfaction and development. But it was stimulating to begin to get an idea of how to invite such a receptivity of mind . . . and how it felt consciously and at will to "tune in" on the mind of another!

Many experiences commenced coming to me, once I had hit upon the "technique" of mental operation required to exercise the extrasensory perceptive faculties we all possess. But, while they all held a significance for me, they could mean nothing to others, since the impressions I had received had not been accomplished under scientifically controlled conditions.

For this reason, I welcomed the opportunity to test out these powers of mind with Sir Hubert Wilkins. Experiments, such as we contemplated in long-distance telepathy, could be conducted under scientific observation. Furthermore, Wilkins and I would be separated by more than three thousand miles, with short-wave radio and airmail the only possible means of physical communication.

This made test conditions ideal, since I could turn over copies of each night's recordings to competent scientific witnesses shortly after setting down my impressions, while days and weeks must pass before any physical contact could ordinarily be made with Sir Hubert, somewhere in the Far North, for checking against his log and diary on that particular day's activities of the search expedition itself.

Dr. Gardner Murphy, then head of the Parapsychology Department of Columbia University, kindly consented to act as an observer after the experiments had been under way about a month, and instructed me to mail him each night a copy of my recorded impressions immediately after their reception, so that they would be protected by government postmark, and would carry their own proof that they were out of my hands and in his, for registering and filing, the very next morning.

Two other scientists had taken an interest from the start in these experiments, Drs. A. E. Strath-Gordon and Henry S. W. Hardwicke, both of whom kept a careful check on developments throughout. In addition, Wilkins and I had a mutual friend and skeptic in the person of Samuel Emery, resident member of the City Club of New York, who was furnished complete copies of all recorded impressions on the same basis as Dr. Murphy.

As soon as Wilkins took off from New York for the Far North, going up by easy stages to gather his crew and equipment in Canada on the way, I began sitting for impressions on schedule. It was my desire to determine, if possible, whether I was getting a significant degree of accurate impressions which could be checked with Wilkins before he got beyond the reaches of civilization, and into the hazardous part of his adventure.

The first night, October 25, 1937, I knew Wilkins was somewhere in Canada. But I was entirely unfamiliar with the northwestern country, having never been there. I was also lacking in knowledge of longitude and latitude. Technically I knew nothing of planes or their construction, although greatly interested in the romance and drama of flying. I was glad for this ignorance, because I had discovered, as Jacques Romano had declared, that "the more one consciously knows or thinks he knows about people or conditions, the more his impressions from the minds of others are going to be colored."

For example, had I ever been in the Far North country, it would more easily have incited my imaginative faculties to action, causing them, in many instances, to embroider the mental pictures coming through, adding elements to them not related to the actual impressions at all. This behavior, in which a genuine telepathic impression becomes fused, in reception, with thoughts already in one's own mind, is called "coloring."

Romano has obtained his best impressions of most persons at their first meeting. After he has met a person several times, he feels he knows too much about him to tell him anything evidential that has not been colored by his previously acquired knowledge.

In the case of Wilkins, however, I was fortunately completely in the dark concerning the nature of the locale he moved in, and of each day's activities. This helped me greatly in maintaining a relaxed mental attitude, freed of anxiety or the tensing responsibility of attempting to weed out any "coloring" thoughts. My own mind, having no background of previous experience or knowledge of his doings to build upon, could therefore suggest little or nothing.

I had confidence, under these circumstances, that whatever thoughts, mental pictures, or strong-feeling impressions came to me, in this receptive state, were actually not products of my mind, but were really being received from without.

And, the very first night, I was electrified at getting the sensation that a circuit had been closed between Sir Hubert Wilkins' mind and my own!

It is a feeling that any investigator will have to get for himself, being one which I find totally impossible to describe. My conscious mind, which is always combative, tried to tell me that this feeling was just my imagination, my own emotions. But I had never felt this way before, and, with the sensation, came a rush of impressions which I commenced to record in a notepad on my desk.

It all did seem incredible, as I wrote, since one impression followed another, almost without a break . . . as though I were a Western Union operator, taking down a series of messages being telegraphed to me by some distant sender.

In place of my expected straining to try to receive impressions of Wilkins' numerical location, I was getting vivid mental pictures of happenings and conditions which I felt applied to him. And there danced before my inner consciousness, on what I have chosen to call "my mental motion-picture screen," a maze of kaleidoscopic scenes.

There came to me the feeling that I was somehow linked with Wilkins' subconscious, and that these scenes were his mental recordings of events that had happened to him throughout his life!

How was I ever going to be able to develop any selectivity, with this merry-go-round of feelings and mental pictures pressing in on me? How could I tune out Wilkins' *past* experiences, and focus my mind's attention on his thoughts of the *present* moment, which he might be trying to send me . . . or determine what had happened to him *this particular day,* out of *all* the days he had lived and experienced?

Then there came to me the thought that man's mind must operate on somewhat the same principle as a radio. There must be a means of selective control. Why could I not instruct my inner mind, through the power of suggestion, to "secure for me the knowledge of what *has* happened to Sir Hubert Wilkins *this day,* or *is* happening to him *now!*"

The instant I strongly willed this thought, to my surprise and relief, all this kaleidoscopic swirl of mental pictures and feelings left my consciousness. It was as though I had been trying to dial a certain station with my radio, and had been hearing any number of conflicting stations in the background. Then, as I made my tuning more selective, and set my dial at the very spot where this station alone should come in, everything else faded out!

On concluding my first night's recording of impressions, my conscious mind—now back in command—cast doubts upon the whole procedure. "It is fantastic for you to believe that these impressions you have listed actually came from the mind of Wilkins," it said to me. "You will probably find that they are all wrong. Why, to have picked up as many impressions as easily and speedily as this is unthinkable. You've let your emotions run away with you. Your desire to achieve results has stimulated your imagination, and tricked you into believing you were getting impressions!"

One's conscious mind always talks in this fashion, for it has been trained to depend upon the testimony of the five physical senses, and to rely only upon what it thinks it knows from the standpoint of past education and experience. To make the conscious mind passive, to eliminate it entirely during an attempted receiving of telepathic impressions, is a difficult job because the conscious mind constantly tries to assert itself.

I was aware, from the start, that the telepathic experiments being undertaken by Sir Hubert Wilkins and myself were of a distinctly pioneering nature. No other two humans, to my knowledge, had ever attempted any such test, on a regular week to week basis, and operating at such a distance. I could not let myself be too impressed with this fact, because my feeling of responsibility would develop a tenseness that would destroy receptivity.

Wilkins had introduced me to Reginald Iversen, chief radio operator for the *New York Times*, with the thought that our experiments might be expedited by my turning over to Iversen copies of outstanding impressions, and letting him transmit them to Wilkins via short wave, so he could check for accuracy against his log and diary. However, as Iversen later testified, magnetic and sunspot conditions were so bad during the entire five months' period that he was able to contact Wilkins but *thirteen times,* and then only for short intervals.

This, as it turned out, was an act of providence insofar as the tests were concerned. For, with Wilkins almost entirely cut off from civilization, no scientific observer could offer the explanation that I, as the receiver, had had access to Wilkins through any possible physical medium.

On making typewritten copies of each night's notated impressions, I would dispatch one carbon immediately to Wilkins, airmail, sending it to whatever point in Canada, Alaska, or the Far North was his last known address. Eventually I was certain this material would be forwarded to him, or he would pick it up at some outpost, even though my letters might have accumulated for several weeks or months. In due course of time, Wilkins would then have an opportunity to check my recordings against his log or recollections of real happenings, and report back to me.

I naturally awaited first word from Wilkins in a state of considerable apprehension. I had made a mental resolve to record all impressions which might come to me, during the period of each "sitting," regardless of how ridiculous or impossible they might seem to my conscious mind. To persist in this resolution, despite the resistance of the conscious mind when impressions seemed to violate all known or presumed facts or conditions, required great fixedness of purpose and exercise of will power. Those who may try such experimentation will soon meet with the same obstructing influence. But it must be faced and overcome, or no genuine impressions can be brought through.

It is fatal to attempt to modify a feeling or mental-picture impression in keeping with what appears to be reason or logic. Reason and logic are departments of mind concerned only with that which is considered to be definitely known or accepted by the individual, having no relation to the phenomenon that is bringing hitherto unknown facts into one's stream of consciousness.

After each one of my sessions, I would examine my recorded impressions, giving my reason and logic full play. Invariably, I would be distressed by the refusal of my conscious mind to accept many of the impressions as possible or correct. This would amount to actual physical and mental suffering. I would reproach myself for ever having jotted down such thoughts or ideas as coming from the mind of Wilkins, and what he might be experiencing. Doubt and wonderment would assail me so strongly, particularly in the first weeks of the experiments, before I had had any report from Wilkins on our work, that making my mind truly receptive at each appointed time was an ordeal.

My nerve tension may easily be imagined when, on the morning of Wednesday, November 10th, 1937, I dropped in at the City Club to get Samuel Emery's reaction to my previous night's test, only to find a letter from Sir Hubert awaiting me, enclosing the checked copies of my first two weeks' impressions which I had mailed to him at Winnipeg.

The letter from Wilkins was dated Sunday night, November 7th, and read, in part, as follows:

Dear Sherman:

> Goodness only knows when I will get around to keeping an appointment with you. Not until I leave here, I expect, for while people are doing so many things for us, I feel I should accept their invitations and go out. That means that, every day, at lunch and at dinner and late into the night, I am talking and answering questions.
>
> The job of getting all my work done keeps me busy all day—so much of it practical work needing my personal attention, so that even if I had a private secretary, I would have little time to use one except at night—and then secretaries don't usually work.
>
> Therefore, I am taking the shorter way to give you notes on the "tests." I am notating on my copy. I will return them to you so that you will see how far they check, and I will be glad if you will save them for me, and I will pick them up upon my return . . .

I was reading this letter to Mr. Emery at the time, and when I got to this point, I felt mortified and heartsick. Wilkins had stated that he had not been able to keep any appointments with me, and yet I had recorded impressions each night, on regular schedule, having had the mental feeling that I had been in actual contact with his mind. This meant, no doubt, that what I had received, up to this point, had only been the result of a fertile imagination. I had been tricked by the operations of my own consciousness.

Mr. Emery, good friend that he was, was hardboiled in his attitude toward these "telepathic experiments," having declared: "I believe little—but I deny nothing." He now voiced what I could not bring myself to say.

"Well, I guess if Wilkins didn't have time to work with you, these impressions aren't of any value. Of course, I don't want to throw any cold water on what you're trying to do, but I haven't much faith you'll get results anyway. It doesn't make sense to me."

I felt, at that moment, as though Mr. Emery were right. I had always tried to be my own severest critic and taskmaster. I did not want to be "kidding myself" about the results of any research undertaken by me, having encountered too much evidence of self-delusion in others.

It required an effort for me to return to Wilkins' letter, and face what I thought I would find there—a declaration that, of course, in the light of circumstances, what I had already recorded had no significance. But, to my amazement, I saw that Wilkins had written:

> . . . You evidently have picked up quite a lot of *thought forms*. Strong thoughts emitted during the day, and some of which I would, if I had had time, have tried to pass on to you at night.
>
> I now often direct these thoughts at any time of the day when I think of it, for I *believe that the thought form does not necessarily fade with its first "spread," but keeps revolving in our atmosphere so that a sensitive mind may pick up the form some hours or even years after it has been emitted.*
>
> The impressions you have had so far seem to contain too many *approximately correct* to have been "guess work" on the part of your Conscious mind—and I believe that when we get around to it, we should have some interesting results . . .

By this time my entire body was tingling with a thrill-sensation unlike any I had ever experienced. Then I *had* been in touch with Wilkins' mind after all—whether he had been keeping his mental appointments with me or not. In some miraculous way, not yet known by me, contact had been made between Sir Hubert's subconscious and mine . . . and I had become aware of his thoughts, of what had been happening to him!

I now feverishly examined my recorded impressions, and Wilkins' notations opposite each recording. Samuel Emery listened as I read the results of my first night's test.

"Well, what do you think?" I asked him, when I had finished.

"Very interesting!" he admitted. "But it will take more than this to convince science—or even convince *me!* You will have to get a sufficient number of correct impressions, continuously and over a long enough period of time, to establish the genuineness of this mental phenomenon beyond any shadow of doubt."

An examination of succeeding nights that Wilkins had reported on gave evidence of the same consistency of results. I now was enabled to return to the experiments greatly heartened, and with new zest for the work.

But, on Armistice Night, November 11th, when I had finished the "sitting," I left my study and took my recorded impressions out to my wife.

"Martha," said I, "I'm afraid I've been completely off the track tonight. I was reading in the papers, earlier this evening, of Armistice observances in Washington and other parts of the country. In some way, my conscious mind has interfered with the impressions coming through, and has caused me to mix up Wilkins with Armistice doings. Now, my common sense tells me that he wouldn't be involved in such affairs. He's too busy with his expedition activities."

"Well, you can't help it. You've put down just what came to you—what seemed to be genuine impressions at the time," said Martha, "so I wouldn't worry about it. You've found out, before, when you've heard from Wilkins, that you were much more nearly correct than you thought."

But the dissatisfaction with myself remained. My mind would go back to that night, as time went on, and I tried more and more to guard against any possible intrusion of "coloring" thoughts.

Then, finally, came a large envelope from Wilkins, containing a batch of my recorded impressions, with his usual specific notations. I dreaded to look at the pages of the Armistice material. A nerve reaction hit me which I felt in the pit of my stomach. It was something like the sensation everyone experiences when he is expecting "bad news," or is steeling himself to "hear the worst."

I was so conscientious in my desire to get authentic results that it was hard to bear any thoughts of uncertainty concerning my impressions. The fact that many of them had already proven to be accurate had only served to increase my sense of responsibility, and my resolution to improve on what had gone before.

In this state of mind, I was totally unprepared for the shock that came to me when I looked at Wilkins' report, in which he stated that I had scored my most "significant hits" yet on Armistice Night!

My first reaction was one of profound relief. I had developed an inward tenseness which had affected my digestion. But now I could physically relax, and breathe more easily.

It seemed even more unbelievable to me, as I reviewed matters, and considered that Wilkins had *not intended to participate* in any Armistic ceremonies. Had it not been for the bad weather, compelling his plane to return and land at Regina, he would not have been "roped in," as I recorded it, on any Armistice observance! This was not a *planned experience* in Wilkins' mind—nothing that he knew about ahead of time—and, therefore, an exceedingly good test impression to have picked up.

But, that Armistice Night, as I had sat, my mind blanked of any conscious thoughts, and receptive, I had heard, in what I might call "my inner ear," Wilkins' voice making an address. I quickly jotted down what I felt to be the highlights of this talk, since I could sense other impressions crowding in, waiting to pour through on me. And, when I had finished, I noted: "These thoughts running through your mind *if not actually uttered in Armistice address.*"

To think that Wilkins had made such a talk, *expressing these very thoughts,* in practically the *same words,* at the "breakfast farewell occasion," before taking

# What Happened to Me

off from Winnipeg! Not only that, but the fact that I had "seen" what took place there, in my mind's eye, so clearly that I had described the pinning of a medal on his coat lapel!

And, what to me was even more astonishing, that I should have received "mental picture impressions" of the Armistice Ball held that evening, which was perhaps taking place even while I was sitting! I recalled how I had said to my wife:

"Imagine my recording that Wilkins attended this affair—*in evening dress!* I'm sure he didn't go north with any such clothes. And yet, I seemed to see him, at this important event—so attired! If I hadn't determined to record every impression that comes to me during each period, I certainly would have rejected this one!"

And yet, now I was confronted with the evidence that even this impression was startlingly correct!

"My appearance at this affair was made possible *by the loan to me of evening dress,*" Wilkins had reported. And when he had finally returned to New York, he added: "This was the *only* time in the five months I was away that I wore evening dress. That you picked up this impression on that *one* occasion is indeed remarkable."

I found myself in the unique position of being a participant in these telepathic experiments, and yet able to assume the role of an outside observer when not actively engaged in the tests, analyzing what had occurred with my conscious mind as would anyone else interested.

Today, after studying the hundreds of impressions recorded by me during the test period, the results obtained on Armistice Night, November 11th, 1937, stand out as among the most surprising of all. And not the least of the true impressions received that night was the one about the "charcoal likeness" of Sir Hubert Wilkins, described by me, stating that he was pleased at seeing it.

When I tried to compute the mathematical chances of my ever having hit upon a "charcoal likeness" by guess work, it made my brain spin. That Wilkins actually was able to testify that a charcoal likeness of himself had been printed in Canadian newspapers *that very day* was an astounding confirmation that I had indeed picked up his thoughts concerning it!

As I had become more accustomed to recording my impressions on this schedule of three nights a week, I had been asked by Dr. Strath-Gordon if I thought it would inhibit my work for someone to be seated in the study with me during the period. I told him I did not know whether this would make me self-conscious and cause me to try to force impressions, or not. So, on two different nights, as an added experiment, not only Dr. Strath-Gordon was present, but his wife, Erica, and my wife, Martha. They all sat on the sofa, and I sat, with my back to them, facing the wall.

I had discovered that darkness was an aid to good reception. This was so because the objects I saw about me in the room, with the lights on, became mental pictures registered in my brain, interfering with impressions I was receiving which were taking-on picture-form in my consciousness.

It was possible to receive impressions with the lights full on, and what I was doing in no way resembled a trance or anything spiritualistic. But I was interested in most effective results, and so I usually switched the lights out when I was ready for a test period. I kept an electric torch by my side which I would turn on when I had an impression to record in my notebook, and I then turned it off again the instant my writing was finished.

On several occasions, delayed downtown past my scheduled time, I would be seated on the subway train, enroute home, at the time of my appointment. I would scribble on the back of an envelope impressions coming to my mind, and later transfer them to my notebook. The roar and rush of the subway, and the constant shift and shuffle of the riders, apparently had no effect—for even these impressions were found to have the same degree of accuracy.

I missed completely on quite a few impressions, of course, and was partially right on a fair number of others, but my percentage of "direct hits," over the entire five months, was exceedingly high.

On the night of December 7, 1937, an unusual thing happened. I had no sooner seated myself in my study, at the appointed hour, and turned off the light, than I suddenly saw a fire flaring up against the black void of my inner mind. My nerves instantly prickled with excitement, and I commenced to record:

*Don't know why, but I seem to see crackling fire shining out in darkness of Aklavik—get a definite fire impression as though house burning—you can see it from your location on ice—I first thought it fire on ice near your tent, but impression persists it is white house burning and quite a crowd gathered around it—people running or hurrying toward flames—bitter cold—stiff breeze blowing . . .*

It was not until February 14, 1938, that I received any check on this impression from Wilkins, who had been beyond the reach of airmail all this time, and my envelopes containing recordings had piled up, waiting for him, at Aklavik.

I noted that I had missed getting a sensing of Wilkins' having left Aklavik, and had placed him there when the fire occurred. Actually, as he testified, Wilkins was in Point Barrow on that night.

The fact that I had not sensed a change of location was interesting to me. Time or space apparently does not exist in telepathic communication. I had made contact with Wilkins' mind just as easily at Point Barrow as I had at Aklavik, but I had not felt the fact of his having physically moved from one point to another.

Yet, as Wilkins told me later, allowing for the difference in time, he was seeing the fire taking place in Point Barrow, and I was writing down my impression of it in New York City at the *same moment!*

This meant that, in some way which cannot as yet be entirely explained, I was attuned to Wilkins' mind, seeing and feeling, simultaneously, what *he* was seeing and feeling!

## What Happened to Me

Most of my impressions, on any night, covered that day's events, being related to activities still fresh in Wilkins' consciousness. This is the most striking instance of a telepathic pick-up of what was in Sir Hubert's mind at the same moment as that mind was engaged in observing the external happening.

As these tests continued, the interest of Reginald Iversen, chief radio operator for the *New York Times,* grew. He finally asked me to let him see Wilkins' report on some two and a half months' listing of impressions which I had just received by airmail from Aklavik.

"Why don't you and Mrs. Iversen come out to our apartment for dinner next Monday evening," I invited, "and we'll go over the impressions together."

"Fine!" Iversen accepted, "but Monday is one of the nights you sit for impressions. That's something I'd like to see. Do you think you could tune in on Wilkins with us present?"

"I'll be glad to try it," I promised, and the date was set for Monday evening, February 21, 1938.

Sir Hubert Wilkins had gone, earlier that month, to Edmonton, Alberta, Canada, to secure a new engine for his plane, following the "crankcase" accident, which I had *foreseen*. He had only recently returned to Aklavik.

As I sat for impressions that particular night, Mr. and Mrs. Iversen and my wife were in the study with me. They remained, quietly observant, on the sofa, until the period was finished. Then I turned to them, notebook in hand, and read the impressions that I had just recorded.

Strangely enough, on this night of all nights, I had the feeling that Wilkins was trying to get in touch with Iversen by short-wave radio, and I had written:

*Think you* (Wilkins) *would like to get some word through to Iversen if you could reach him before Thursday—wonder if this thought in your mind tonight as you think of me—?*

"No, you're wrong on that impression!" said Iversen, positively. "Wilkins knows that I don't work Tuesdays and Wednesdays. He knows that I'm off duty at the *Times* tonight, Monday, at seven. That means I won't be at the office again until Thursday morning. He never tries to reach me during my off-time."

"Perhaps not," said I, "but the feeling still stays with me that Wilkins has some messages for you, and has been trying to get them through tonight."

Iversen was amused, but unconvinced.

"There's something to this telepathic business all right," he admitted. "You've received too many correct impressions over all this time. But you're wrong about Wilkins trying to radio me tonight. I guess my being here has influenced you to put that down."

The Iversens left some time after 2 A.M., for their home in Astoria, Queens, Long Island.

Next morning, between 10 and 11, an excited Reginald Iversen phoned me.

"Say, I'm down at the *New York Times*—just came here to pick up two radio messages that were received by the night operator from Wilkins last evening, while we were up at your apartment! The operator phoned the house several times last evening, trying to reach me. This is the darnedest thing I ever heard tell of. Sherman, how the deuce do you do it?"

"I don't exactly know myself," was my answer. "All I know is what state of receptivity to take on—and then things happen."

"I'll say they do!" was Iversen's rejoinder. "Didn't you read me some impression that came through last night about the new plane engine?"

"Yes," said I. "Just a minute till I get my notebook." I then read from the originally transcribed impressions. "*Installing of engine has been completed, and testing of it carried out today.*"

"Correct!" cried Iversen. "Listen to this radio message: '*Engine on plane mounted—won't be ready till Thursday*' . . . and that's not all, Sherman . . . you had some impression about Wilkins wirelessing me for equipment. Look back through your notes, and read it to me!"

I thumbed the pages to the date line of February 14th, and read: "*Kenyon seems to want more supplies or parts of equipment . . . you have to delegate New York, through Iversen, to secure some pieces equipment and rush through in relation to repair job on plane's motor . . . also in connection radio . . .*'"

"That's it!" said Iversen. "*Radio!* . . . Listen . . . here's Wilkins' second message which was sent last night: '*Unless you already ordered, hold everything ref: Kenyon's radio receiver. He will probably not get until after return.*'"

"Wait a minute!" I cut in. "I got a follow-up impression on that last night, as well as several nights ago. Don't you remember the impression I read you: '. . . *some member of crew requested you to get some article . . . can't make out what it was*' . . . My impression wasn't as definite last night, but this indicates I was closely tuned to his thought on the radio."

"This beats me!" said Iversen. "At this rate, I'll bet most of those other impressions you got last night turn out to be okay, too!"

Iversen's conjecture proved to be right. The recording of these additional impressions witnessed by the Iversens and my wife ultimately received confirmation from Wilkins.

One of them was particularly unusual. I had recorded:

". . . *someone has skinned hand or finger . . .*" And Wilkins had replied:

"*Dyne had hands spotted with frost 'burns' which blister, or the skin is pulled right off when the hand is pulled away after being frozen to any metal . . .*"

I had actually *felt* the condition of the hands, through my emotions, as I recorded the impression!

March 23rd, 1938, proved to be the date of my last "sitting," and by this time the mental and physical strain of keeping up these tests had seriously impaired my health. I had carried on my regular writing activities, had written several book manuscripts and a musical comedy, during this time—in addition to quite a number of short stories. Looking back, I do not see how I was ever equal to the pressure and volume of the work entailed by these tests, which turned out to be much more prodigious than I had ever conceived at the start.

Living up relentlessly to this schedule required severe mental discipline. And the realization that the routine, as interesting and exciting as it had been, was at last over, brought a sense of great relief to me. It was wonderful to be able to take a vacation from mental things, but, even so, most difficult to do, for the mind had become so highly sensitized that it kept bringing me unsought impressions and unusual mental flashes.

Inwardly, however, there had come to me a new strength of conviction. I knew now, beyond any assailments of doubt, that telepathy was not only a fact, but that its power could be developed and demonstrated.

I knew, also, the dangers of over-exercising the mind, and the penalty exacted of the body by so doing. I would never again put myself to such a test, unless I were freed of other responsibilities. And, even then, I would have to watch carefully that I should not expend too great an amount of physical and mental energy.

I walked the floor many nights in great pain brought on by nervous indigestion and stomach cramps. My system was reacting against the abnormal demands I was making upon it—demands which I now know could have cost me my life.

# 2

# How Thoughts Are Received

WHEN THE RESULTS OF these telepathic experiments were announced in the *Cosmopolitan* Magazine for March, 1939, in an article written by Inez Haynes Irwin, and when Sir Hubert Wilkins and I appeared on the "Strange As It Seems" radio hour, December 12, 1940, in an interview and a dramatization of some of the most outstanding instances of thought transference, we were deluged with thousands of letters, with over 96 per cent of the correspondents asking: "How is telepathy performed? Where can we obtain further reliable information about it?"

The faculty of telepathy has been used by humans in every age, apparently, since the world began.

As Wilkins has reported, he found that certain bush tribes in his native country, Australia, used a form of mind communication.

Alexander Markey declares that the Maoris of New Zealand communicate by means of telepathy today. In his treks through the jungle from one tribe to another, he found that word of his coming had precede him. Upon his arrival, the whole tribe would be assembled to welcome him, although there had been no physical means for this information to reach them.

We are all familiar with the stories which have come out of that mysterious and occult country, India—the land of fakirs—where so-called "supernormal feats" of mind and body are said to be commonplace.

I doubt if this development is as widespread as many exaggerated accounts would indicate, but quite unquestionably there are a number of Hindus who possess great powers of mind.

Certain of these men have spent a life-time practicing concentration. I believe that, had any individual devoted the same amount of time, he also

## How Thoughts Are Received

would be able to demonstrate what is today still considered "remarkable faculties of mind."

While, as Wilkins states, some individuals appear to be more naturally endowed by temperament and mental attitudes to develop a degree of sensitivity to the thoughts of others, it is my conviction that we all possess essentially the same potentialities in consciousness.

It is perfectly apparent that wordly education has nothing to do with these powers. They operate irrespective of cultural or scholastic attainments. In fact, those whose religious or intellectual beliefs are opposed to the possible existence of such mental phenomena seem to close it out of their experience by this very attitude.

I found, in the beginning, that I had to assume that telepathy *could* be performed, in order to develop and maintain the state of receptivity necessary for results.

When I allowed elements of doubt to creep in, it seemed to demagnetize certain inner forces in my mind, and eventually led to confusion and loss of concentration.

You will find, if you are seriously interested in seeing what you can do in the field of telepathy, either for your own enlightenment or for scientific purposes, that you will need to put aside, by an effort of will, all doubts and skepticism.

You will have to accept, for the time being, the belief that telepathy is an established fact. Take it on faith. You cannot be wondering and guessing whether what you are attempting will work or not at the time you are seeking to receive impressions.

To do this will defeat all efforts. Your conscious mind will set up all manner of obstructions, and it also will bring your imagination into play, causing you to color or distort any true impression that might have been received by your subconscious.

As nearly as can be determined, telepathy is a function that operates between the subconscious mind of the sender and the subconscious mind of the receiver.

When a thought is intensely visualized and projected by a sender, it makes such a vivid impression upon the subconscious mind of the receiver that this impression is relayed to the conscious mind; and the receiver, sensing it, is enabled to express the thought in his own words, or to describe the flashing mental picture that comes to him.

Telepathy is basically a simple process. There is nothing mysterious about it, and any fair-minded person, willing to expend the time and effort required, can prove the existence of this power of mind for himself.

Only the average man's own almost complete reliance on the testimony of his five physical senses has kept him from recognizing, developing, and using the higher mental powers long ago.

Telepathy, however, is a faculty of mind in the operation of which you must be able to forget all that you *think* you know about anything. Make your inner mind as clear and passive as a pool of water, so that it will reflect the vaguest shadow, or so that it will react to the tiniest pebble which may cause a ripple upon it.

Permitting this mental pool to be filled with the images of past thoughts or convictions obscures reflection and clouds your vision. You cannot have any of your thoughts lurking in the pool if you hope to secure a true reflection of the thoughts of others in it.

Fortunately for me, in my telepathic work with Wilkins, my mind was filled with other matters during each day. I refused to let it indulge in any wonderment or speculation as to what might be happening to Wilkins. I wanted to approach each appointed time with my mind absolutely clear and fresh, having no knowledge or inkling or suggestion lingering in my conscious mind of what could be happening to Sir Hubert. I was thus able to bring to each sitting a natural enthusiasm, and an eagerness to receive messages from him. This attitude seemed to create in me an *energy* that I needed to get results.

It was as though I had to charge a mysterious battery inside me; to generate a "receiving power"; to take on a certain, indescribable "inner feeling sense" or vibration. Words are dangerous things to use in trying to define states of consciousness, and few words have as yet been coined to convey adequately the proper meaning in setting forth one's experiences in mental phenomena. But I had learned, by degrees, to use a technique for bringing me impressions from without that appeared to work.

I would first relax my physical body, from head to foot, letting go of it, with my conscious mind, completely. This, in itself, is not easy. I then, having made my conscious mind passive, would look *within my mind,* figuratively turning my physical eyesight *inward,* focusing it upon what I call my "mental motion-picture screen." This is an imaginary white screen that hangs in the darkness of inner consciousness, and, upon which, as in the reception of television, are flashed the images that hit my mind from outside sources.

Almost instantly, when this technique had been followed, and I was truly relaxed—physically and mentally, without any sense of strain or attempt at forcing—I would commence to see and feel things. After that I had only to put these mental pictures and sensations into words.

Usually, in my work with Wilkins, I could feel my mind make "contact" with his mind. I could distinctly sense a force, a line, or stream of energy that seemed to be connecting my subconscious with his. On the nights when I felt this the strongest, I got my best results. As time went on, I became certain enough of this condition occasionally to make a note, along with the recorded impressions, that I felt assured of a good evening's results. At other times, when approaching a sitting, mentally and physically tired from a day's activities, I would express my misgivings.

## How Thoughts Are Received

Despite the variation in physical feelings from sitting to sitting, however, I was interested to observe—from a study of the results—that I had been largely successful in throwing off sensations of ill health and tensions of the day, and that the tests appeared little affected by how I might have felt personally.

As long as I could, through an effort of will, "retire from my external world," leaving the consciousness of the body behind, I could apparently receive impressions with the same average consistency.

This fact clearly indicated that exercise of the telepathic faculty depended more upon the ability to develop and maintain the right state of receptivity of mind than anything else.

Years before, prior to the time I had learned how to induce the reception of impressions, I experienced what seemed to be occasions of thought transference while about my ordinary day's activities. What now appear to have been outside impulses would reach my consciousness, infrequently, and were so strong that they would break through, and would force themselves upon my conscious mind's attention. I would then be "made" to think of a certain person, or I would get a decided urge to do something, or get in touch with somebody. But the telepathic contact, in these cases, was seldom recognizable, and I responded as though the thought had been my own.

You, and everyone, have had similar experiences—a telephone call or a letter from some individual who has not been heard from or thought of in months or years, and who, for no reason whatsoever, has come strongly to your mind shortly preceding your receipt of a message from him.

"Why, I was just thinking of you!" is the oft exclaimed expression on an occasion when a person has shown up suddenly and unexpectedly.

But happenings of this nature are at once set down as "coincidences," without any further attempt at explanation.

Mark Twain had several incidents occur in his life which he was able to identify as evidences of what he called "mental telegraphy."

One day, after he had become a world-renowned writer, he hit upon the idea that there should be a good book in his early western adventure with Dan DeQuille, old prospector friend, who lived in Nevada.

On the inspiration of the moment, he wrote Dan a letter, outlining his plan, and telling Dan he would like to collaborate on such a book with him, drawing from his rich memory store of anecdotes and incidents covering this period.

Mark proposed that Dan arrange to come East for a few months, so that they could work on the book together, and promised to pay all expenses.

But, with the letter completed and the envelope sealed and addressed ready for mailing, Mark had a change of heart.

"What am I doing?" he asked himself. "I haven't the time to undertake a job like this now. Guess I'd better hold Dan's letter up until I can get clear of my other work."

And so Mark stuck the letter in a pigeon-hole of his desk.

Some ten days later a friend was visiting Twain, when the mailman arrived, leaving a letter for Mark addressed in handwriting which was strangely and distantly familiar. Mark studied the letter without opening it, and then handed it to the friend and said:

"Here, take this and hold it—while I perform a miracle!"

Mark then proceeded to tell this friend that the letter in his hands contained a message from Old Dan DeQuille, from whom he had not heard in years.

"Dan has gotten the idea," said Mark, "that he and I could write a book together on our prospecting experiences. He's made some notes on what happened to us, and he wants to come East and visit me, if I will pay his expenses, so he can give me his recollections. Now, open the letter and see if this isn't so!"

The friend did as directed, and then exclaimed in amazement.

"You are absolutely right!" he said. "But how did you know?"

Mark smiled and reached into his desk, taking out the letter he had written and addressed to Dan DeQuille.

"Because," he said, "I made this very proposal to Dan almost two weeks ago, and here is my sealed unmailed letter to prove it! Open it and see for yourself!"

The friend followed instructions, and found the contents to be the same in every essential as the letter written by Dan DeQuille to Mark.

"How do you account for this?" the friend asked.

"There is only one way to account for it," said Mark. "In some manner, Dan DeQuille received my thought impressions of this book I wanted to write with him, even though I didn't mail the letter. But it seemed to him that these thoughts were his own idea, so he sat down, and wrote them to me. The moment his letter came to hand it flashed over me what had happened, and I *knew* my thoughts had reached his mind. I consider this a clear case of *thought transference.*"

Most of us have been experiencing somewhat similar telepathic phenomena in our everyday life without realizing it. Or, if we have, it has not been possible for us to bring the function under our conscious control. We have found ourselves incapable of duplicating such an experience at will. All we know is that something unusual has happened, something entirely unpremeditated, and something apparently having nothing to do with our normal mental activity.

This phenomenon has usually taken place when we have fixed our minds on certain situations which are affecting us deeply. Under such stress of emotions, we sometimes discover that we have unconsciously tuned in on the minds of others interested, and have received "uncanny" impressions of conditions and developments beyond the possible knowledge of our five physical senses.

But, for the most part, these occurrences of the kind that happened to Twain are accepted as a matter of fact, the true telepathic impressions being so fused with our own thoughts that we are unaware of the manner in which they came to consciousness.

Your conscious mind is the great barrier to reception. It is well that this is so, for you would have little protection from the thoughts of others were it not.

In sleep, your conscious mind is completely passive and at rest. And during the mysterious process of sleep we have thousands upon thousands of authenticated cases of thought transference.

You have certainly heard variations of the story of the mother who awakened from a sound deep sleep, crying out that she had a vision of her sailor-son drowned at sea. Members of the family try to calm her fears, telling her she has simply experienced a nightmare, that everything is all right.

"Oh, no!" the mother always insists, "I know John has been drowned. I saw him standing before me, soaking wet; I heard him say, 'Goodbye, Mother.' I got the impression he had been washed overboard in a terrible storm. He's dead, I tell you. I can't explain how I know—but I *know!*"

And several days later, word arrives that John has indeed been drowned at sea, and usually it is found that the time of drowning synchronizes with the time of the so-called dream or vision.

Conditions had automatically proven perfect for telepathic communication.

John, in the act of drowning, undergoes the emotional crisis of his life. Sensing the approach of death, he naturally thinks of his nearest and dearest of kin—his mother. His desire to see her again, to get some last message to her, causes his whole being to be charged with this one great yearning thought.

Had it been daytime and his mother up and about the house with her conscious mind occupied with household duties, she might never have received the impression. All that she might have sensed would have been a fleeting uneasy feeling about John, which her conscious mind would have brushed aside.

The message John was sending her would be received by her subconscious mind just the same—but it might not be able to make her aware of it against the conscious mind's resistance.

But, in the sleep state, John's mother was completely receptive to impressions from without her own consciousness.

John's intense thoughts reached her subconscious mind and recorded their images there so positively and distinctly that she was awakened with them still vividly impressed on her consciousness. Her conscious mind, called into activity, had nothing else to focus its attention on at the moment except these impressions—and the carry-over proved instantaneous and complete.

"I can't explain how I know—but I know!" the mother repeated, over and over.

And anyone who has undergone a similar experience can understand and appreciate the unmistakable feeling.

You do not know how you know—but you know! And your conscious mind, under such circumstances, cannot argue you out of it. Neither can anyone else.

A prominent Philadelphian was returning from a trip abroad several years ago. While still two days out of New York, he had a vivid dream in which he saw his son, a young man of twenty, killed in an automobile accident. He saw the car topple over an embankment, and pin him underneath.

Awakened by the shocking impression, he was unable to get back to sleep. At breakfast he related the incident to friends on board, who, naturally, tried to make light of it.

"Something you ate last night," jollied one. "There's nothing to dreams, anyway."

But the father was so worried that he insisted on putting through a ship-to-shore telephone call to his wife in Philadelphia.

"Why, no, dear," said his wife. "I don't know how you ever got such an impression. Of course, Richard's all right. Now put it right out of your mind, and enjoy your trip. I'll meet you at the boat."

"Well, tell Richard to be careful!" urged the father. "I can't understand . . . I can't shake this feeling. But if it hasn't happened, don't let him touch the car till I get home!"

Reassured by his wife, the troubled father hung up the receiver to face the gentle ribbing of his friends.

"You see—there wasn't anything to it. But if that made you feel any better . . . !"

"It didn't much," said the Philadelphian. "That's strange. I'd have sworn . . . it was so real . . . oh, well, maybe you're right . . . maybe there wasn't anything to it."

But despite all concrete evidence to the contrary, this father could not enter into the ship's gaiety. He could hardly wait for the ship to reach port and dock.

On the pier his eyes sought out the form of his wife. As he beheld her, he started.

"My God!" he exclaimed, "I *knew* it! I was right!"

His wife was dressed in black, and heavily veiled. She was weeping.

"Oh, darling!" she said, as he took her in his arms. "It was true. It had happened just as you said . . . but I didn't have the heart to tell you, with you two days away from me—knowing how crushed you'd be! Richard must have been killed just about the time of your dream!"

Experiences such as this occur too often, and are too well authenticated to be dismissed or denied. They are happening to scores and scores of people throughout the world every day. But those of which we hear represent only the outstanding instances in which conditions for thought transmission have been close to ideal.

Obviously, if thoughts can be transmitted by individuals, particularly by those who are under great stress of mind and emotion, and these same thoughts

# How Thoughts Are Received

can be received by others whose minds are in the receptive state of *sleep,* then it should be possible to develop a technique for making the conscious mind so passive that it is capable of receiving similar thoughts in the *waking* state!

I started my early research on this premise, encouraged also by what I regarded as telepathic impressions which had come to me while awake, both sought and unsought.

My problem was to hit upon a method which could accomplish, through practice, a state of relaxation and receptivity equivalent to that attained just before one drops off to sleep—in which state I believe one could make his consciousness subservient and sensitive to thought impressions from the mind or minds of others.

What I really set out to do was to gain conscious control of my subconscious mind—to be able to reach it at will. Once I could do that, it seemed to me, I could make my conscious mind aware of any impression my subconscious mind might be receiving.

Usually there is little planned synchronization between conscious and subconscious minds. You do with your conscious mind what you wish, irrespective of the subconscious. And yet I believe that your subconscious is the more important of your two divisions of mind.

This subconsciousness of yours not only controls and directs the functioning of your different bodily organs, heart, lungs, stomach, and so on, including your involuntary nervous system, but it serves as the memory storehouse for all the reactions you have had to the things that *have* happened and *are* happening to you in life.

More than that, your subconscious contains an apparently unlimited creative power which acts upon the mental pictures of your desires which are given to it by your conscious mind.

Whatever you picture in your mind, if you picture it vividly and earnestly enough, is eventually attracted to you, in your outer world, by this creative power. How this is actually done, science still doesn't know. But the old adage, "as a man thinketh, so is he," is profoundly true.

And what we are now calling "extrasensory perception" is playing its vital part, as it always has, behind the scenes, helping us to set the stage for each new act of our lives.

If this is true, then it is to our best interests to gain such knowledge and control as we can of the powers of mind we possess—powers we are now utilizing, largely through ignorance or disregard, as much *against* as *for* ourselves, dependent upon the nature and character of the thoughts we think!

You've probably said, many times, "I was afraid that was going to happen." This statement was nothing more than a confession that you had visualized over and over in your mind's eye the possibility of something dire occurring—and, lo and behold—it finally *had!*

This was no accident. You had created this unhappy experience for yourself by commanding this creative power to produce it for you—attracting the conditions, circumstances, and people necessary in your external life to bring it about.

It is necessary that you understand a few of these fundamentals of mind operation, if you wish to explore the mysteries of your own inner mind—your own real self.

The duty of your conscious mind is to deal with this external world in which you live. It is your reasoning, guessing, wondering, doubting, calculating, assimilating mind. It is bounded by your five physical senses—sight, hearing, taste, touch, and smell.

Your conscious mind, during every waking moment, is reporting to your subconscious or inner mind, in the form of mental pictures, your emotional reaction to everything that is happening to you in this outer world.

These pictures continue to live in consciousness, spurring the creative power on to materialize in your physical life the things you have visualized.

Your subconscious mind has no reasoning power. Therefore, if you permit mental pictures dominated by fear and worry and other destructive emotions to reach it, your subconscious accepts these pictures as what you want, and starts to work, as a builder acting upon the blueprint of an architect, to attract the elements required. Sooner or later, unless you change the nature of these thought pictures, you will find them reproduced in life for you.

This was a discovery I made in the course of my research and experimentation in mind—but the discovery was not original with me. Thousands of humans throughout all ages have been utilizing this law of mind, and the creative power they possess in consciousness—through a properly developed faculty of visualization.

The law, and the method of operating this law, bring one close to a knowledge of how to perform telepathy—in addition to enabling a person to accomplish the things in life he most desires.

If the average individual, who has been prone to scoff at reported higher mental powers, would only determine to assume an open, investigative attitude of mind, he could soon prove the existence of these powers in himself, and could develop them for his own practical, everyday benefit.

I am now going to tell you the exact steps to be taken for gaining *conscious will* control of your inner mind; in other words, how to operate the law of the subconscious.

To the degree that you are willing and able faithfully and conscientiously to devote the time and effort necessary to practice and perfect these steps—*just* to that degree will you get satisfactory, evidential results.

Training your mind is not unlike developing the muscles in your body. If you continue the proper arm and leg exercises, day after day, there will be a noticeable increase in the strength and size of those arm and leg muscles at the end of a six month's or year's period.

### How Thoughts Are Received

Application of this mental law over a similar length of time will accomplish wonders in the life of anyone who sincerely and earnestly undertakes it.

Relaxation of your physical body is the first step.

Few humans really know how to relax—to relieve themselves of the nerve tensions of the day. And yet all physical disturbances must be put to rest before true concentration can begin.

Take as comfortable a position in an easy chair as you can, or stretch out upon the divan or bed.

Fix the attention of your conscious mind upon your right or left leg. By an exercise of will, lift this leg and hold it suspended, momentarily. Then let go of the leg with your mind, and let it drop as though it is dropping off your body—as though it is no longer a part of you.

Do the same with your other leg, withdrawing the attention of your conscious mind from it as you do so.

Now, in your mind's eye, see and feel your conscious mind traveling up your body, letting go its hold upon successive portions of your body in the process.

When the attention of your conscious mind reaches your chest and arms, lift each arm and let it drop "off the body."

You are imaginatively, as well as literally, withdrawing the attention of your conscious mind from your entire body, giving it over to the chair or divan or bed to support—temporarily laying the body aside.

Through exercising such a method, you are cutting your conscious mind off from all bodily tensions, releasing muscular and nervous tie-ups.

Many persons, attempting this technique for the first time, find the relief so great that they fall asleep.

Your head is the area in which the real entity or self resides. The instant your conscious mind's attention is centered in your head, there will come over you a sensation of existing only in consciousness, devoid of your body.

The next step is to make your conscious mind passive. Having freed your mind of its awareness of the body, you will discover that you are in a field of thought in which all kinds of distorted images and impressions which you cannot control seem to run helter-skelter, back and forth across the threshold of your mind. Perhaps the outstanding fears and worries of the particular day or your life as a whole now commence to press in upon you—the very things you would like to forget, and to put out of consciousness. But now, with your body relaxed and temporarily out of the way, these emotionalized thought forms come to plague you.

How on earth can you get rid of them?

You can get rid of them by using an extremely simple device which does not require months of training in so called courses of occult mind development.

Visualize, in your mind's eye, that "blank, white motion-picture screen" of which I have spoken. See it hanging in the darkness of your inner consciousness, and fix your attention upon it. What you have needed, in order to eliminate all irrelevant and confusing thought impressions, is a focal point; and this gives it to you.

With this blank, white motion-picture screen established in consciousness, imagine that you are the projection machine, and that, until you get ready to throw a picture on that screen, *no picture can appear!*

You will find, after a little practice, that you can hold this mental screen blank; that all other hitherto uncontrolled impressions in your mind have been eliminated!

Now you are prepared—with your physical body relaxed, and your conscious mind passive—for the third and last step.

The time has come for you to visualize—in picture form—what you desire to achieve in life.

See yourself, in your mind's eye, as having already attained what you wish, what you are fervently willing to put forth every effort to become or to have.

This mental picture, thrown on your blank, white motion-picture screen, is the "blueprint designed by the architect" which you are turning over to the builder—this creative power within your subconscious—for execution.

If the blueprint has been perfectly conceived, and clearly and confidently projected on the screen—you can depend on the fact that your creative power will reproduce it, according to specifications, in your external life.

The degree of time required for this to happen will be in proportion to the earnestness of your desire and your willingness to work to bring what you have visualized to pass.

You must not permit any fears and worries of possible failure to crowd in, and alter your picture—otherwise you will create flaws in the "blueprint"—and these flaws will appear in the finished job.

Remember—your subconscious mind possesses no reasoning power. It will not check you when you give it a wrong mental picture. It accepts whatever you visualize for it to do, and it sets out, in a manner which can still not be comprehended, to attract to you the elements needed to materialize what you have pictured for yourself.

The operation of this creative power within seems to be infallible and mathematical.

A certain definite mental attitude, maintained over a period of time, has been shown to beget a certain definite result.

You cannot "force" anything to happen mentally. When you attempt to force your mental pictures to come true, it is a sign that you are over-anxious and afraid that what you desire is not going to come to pass.

In the same way it is not possible for you to "force" an image when attempting to receive telepathic impressions. Mental straining to get results will call your imagination into play, and any genuine impressions that you may receive will be so colored and distorted as to be unrecognizable in consciousness.

This method or formula for reaching your subconscious mind at will, which I have given you—relaxation of your physical body *plus* passivity of your conscious mind *plus* fixation of your attention upon the one thing in life you most desire—*can also be used for carrying on experiments in extrasensory perception, since it gives you control over these faculties of mind.*

The only difference in procedure is to *refrain from visualizing,* when you have relaxed your physical body, and made your conscious mind passive.

Next, keeping your blank, white motion-picture screen in your mind's eye, *fix your attention upon the individual from whom you hope to receive telepathic impressions!*

If you find it difficult actually to *see* this screen in consciousness, you can obtain the *same* results through your ability to *feel* that it is there. All you need is this *focal point* of attention as the *center* through which thought impressions are to come.

Your attitude must be impersonal, and your own emotions must be stilled. You can, however, permit a *deep yearning* to receive the expected impressions. But your inner consciousness must be freed of any feelings of your own, so that it can be receptive to receiving and feeling sensations which apparently accompany the thought waves or impulses from the mind of the sender.

For instance, one night during the experiments with Wilkins (February 21, 1938) I recorded: "Someone has had toothache—sore condition of mouth."

While in this receptive state of mind, I had suddenly received a severe sensation of toothache, and I could tell that it was a definite impression from without.

Wilkins later confirmed: *"I had tooth filled the evening before I left Edmonton. It was still tender and jumped each time I trod heavily."*

This was the only time in the five months he was away that he had any tooth trouble, a fact which placed this impression far beyond the bounds of "coincidence."

Thought impressions will come to you in one of three forms, or in variations of all of them. You will either "see," or "feel," or "know" what the sender is concentrating upon, and you will find it necessary, in each instance, to interpret in your own words what you sense.

There is no way by which you can determine or control in which of these three ways any given impression is going to reach your consciousness.

Your ability to perceive the fleeting telepathic pictures as they strike your consciousness, and appear momentarily on this motion-picture screen or area where you imagine it to be, depends upon the degree of "inner mental alertness" you can maintain.

You will find that this method of concentration greatly increases your sensitivity. With your conscious mind removed from the awareness of physical nerve reactions and disturbances, and its attention turned inward, it now reveals a surprising ability to detect knowledge, and to tune in on conditions which exist beyond the reaches of your five physical senses.

A sincere, unprejudiced, persevering approach to the development of these higher powers of mind, balanced by the exercise of continued faith and patience in the face of possible repeated failures at first, will ultimately bring you evidential results—as it has to many others.

The phenomenon of extrasensory perception exists, and can be demonstrated. I am not, however, the least bit interested in the performance of extrasensory perception for exhibition purposes.

The end and aim of what research I have done and am doing is acquiring more and more knowledge of how to control and apply these higher mental powers to the solving of practical, everyday problems!

And you, when you have proven the existence of these powers in your own life experience, will not wish to make a parlor game of them. You will instantly sense their deeper, more personal value. And their unfoldment will have enabled you to discover an entirely new world inside yourself—a world of immeasurably greater opportunity for development and advancement.

# 3

## The Part Emotions Play

SCIENTISTS, THROUGH RECENT EXPERIMENTS, are demonstrating that your intelligence is linked with electrical energy in your brain. There is a highly sensitized instrument, known as an electro-encephalograph, which records the electrical fluctuation of the brain after the currents have been amplified more than a *billion* times.

These scientists tell us that electricity generated by the brain is infinitesimal. In a normal person the fluctuating potentials amount to one fifty-millionth of a volt. Enough current to light an ordinary five-watt lamp is said to require more than five times the combined electrical output of the brains of the world's population.

That may be, but the amount of electrical energy in the brain of individual man, if indeed, electrical energy *is* the transmitting medium, is apparently sufficient to send a thought from his subconscious mind to that of another, entirely around the world.

Perhaps Science will discover, as it continues its serious investigation of the little-known elements of human consciousness, that the subtleties of electrical energy, as they apply to the brain, are many, as well as amazingly potent, and that the ether, in which radio waves travel so effectively, is an equally good conductor of *thought waves!*

No one can state positively, as yet, remember, that this is the medium by which thoughts are transmitted. But it can form an understandable basis for theorizing.

It may well be that we all exist in a vast sea of consciousness, and that particles of our thought, when properly magnetized, can make instant union, forming lines of force which carry messages from one segmented mind to another.

We do not know, but we can say, with a substantial degree of certainty, that man's consciousness is electrical in manifestation, and that thoughts appear to have a rate and character of vibration determined by the nature of an individual's emotional reaction to external experience.

As you know, the things that are happening to you in this outer world reach the brain through one or more of your five physical senses.

Your reactions to these happenings take the form of emotionalized mental pictures, which fuse themselves with *similar* mental pictures of *past* happenings, already stored in consciousness, and now called back into your *present* moment to form your idea or feeling at the instant.

This idea or feeling has come into being through a discharge of electrical energy automatically set up in your brain, touched off by the nerve impulses reaching it over your channels of sense.

Thought waves or lines of force, thus created by this discharge of electrical energy, are broadcast into space without your realizing it. The intensity of this discharge depends entirely upon the intensity of the emotion behind the thought or feeling.

It naturally follows that the more intense the external experience, the more vivid will be the emotional reaction in the form of mental pictures sent to the brain. This results in a correspondingly more violent discharge of electrical energy, and more strongly vibrating thought waves or lines of force.

For this reason, outstanding and tragic events, having a profound effect upon the individual, are most likely to reach the consciousness of a close friend or relative miles away, just as broadcasts from a *high-powered* broadcasting station are most easily picked up by all radio receiving sets.

I found, for instance, in my telepathic tests with Sir Hubert Wilkins—as he himself observed—that:

> You seem to get most, if not all the *very strong* thoughts I express throughout the day, and sense the *vivid* conditions, even though I am generally unable to get down to concentration at the appointed time . . .

Our experiments demonstrated conclusively that: *The degree of intensity of one's emotional reaction to external experience determines the intensity of the thought force projected.*

Whatever impressed Wilkins strongly, reached my consciousness with much greater ease and vividness. This struck us both as significant and led to the conviction that: *Human emotions are the batteries which generate the power, according to their intensity, that is put behind the electrical currents of the brain.*

I noted that thought impressions were sensed by me at two points in my body—in the brain and in the solar plexus.

I would get a nerve reaction in the "pit of my stomach," which I came to realize always accompanied a genuine telepathic communication.

# The Part Emotions Play

It is a sensation that any experimenter must experience for himself. It approximates the feeling one sometimes gets when he suffers sudden shock, or when, for no reason that he can explain, he becomes anxious or apprehensive over something.

This state of "inner nerve excitement" arrives simultaneously with the receiving of a mental picture, or of feeling impressions in the brain; and it remains until you have relieved it by recording what has "come through," and "let go of it."

Each thought seems to carry its associated "emotionalized vibration" with it, and your body, as the receiving set, senses the same emotion over its "nerve network."

Your solar plexus is the great "nerve center" of the body. It has been called by many students of the mind, "the second brain."

Here is located the seat of your emotions. When you are disturbed in this area, you are disturbed all over. You must learn, therefore, to be responsive to your emotions, but to maintain control over them at all times.

Your emotional system is an essential factor in the performance of any and all forms of extrasensory perception. If it is not governed properly, it can ruin not only genuine results—but your health as well.

Dr. J. B. Rhine, of Duke University, originator of the ESP card system (containing five symbols—a cross, a circle, a square, a triangle, and wavy lines) for testing the existence of extrasensory perception, has done important pioneering work. But, he has endeavored to *eliminate,* insofar as is possible, the *emotional factor.*

A subject is seated at a desk, with his mind receptive, and is asked to record such mental pictures or impressions of card symbols as appear to come to him from the mind of a sender, perhaps located in an adjoining room, or across the campus, or in another town.

The variations of this system are many. Sometimes no mind knows the order of fifty or a hundred cards, shuffled in a pack. The subject is requested to fix his mind on the pack, and to endeavor to determine the rotation in which the five symbols fall, from the bottom up or down, as the case may be.

This procedure is to check the possible existence of clairvoyant powers as distinguished from telepathic.

The results obtained by Dr. J. B. Rhine are said to demonstrate conclusively the existence of extrasensory perception. His supporters hold that the function of telepathy has been definitely established.

From my own experience, I would say that Dr. Rhine, and others who have patterned their research after his, have gone out to prove the existence of these powers of mind the hardest possible way.

While I have worked with ESP cards, and have obtained above chance results, the experiments have not impressed or intrigued me.

How can I get excited over five symbols that are not related in any way to my emotional system or to the emotions of the sender?

When I was sitting, night after night, making my mind receptive, trying to contact the thoughts in the mind of Sir Hubert, there was always the possibility of some highly emotionalized mental pictures of actual experiences coming to me.

I might tune in on these experiences at any point, and then work back or forward from the point of contact until the complete impression had been brought in.

Let us say that Sir Hubert had flown to Edmonton (which he did at one time), and, while there, had stepped off the curb, and had been knocked down by an automobile (which, fortunately! did not happen).

This accident would have been strongly registered in his consciousness, because of his emotional reaction to it. He would not have to be thinking of it consciously at our appointed time, but, due to the intensity of the vibrations he had set up in his consciousness concerning the happening, I would have been very apt to have pulled in impressions on it.

Events, after they happen, apparently exist in a *straight line of time* in consciousness. This is important to observe. It is the only manner in which I can describe this phenomenon in words. Each *line of time,* as it relates to any incident or human experience, has a *peak emotional moment* in it. The *"peak emotional moment"* in the hypothetical incident of Wilkins' auto accident would have been in the instant he was struck by the car.

A "peak moment" of an entirely different kind is that moment at which you throw a stone into a pool of water. The ripples it sends out are reactions to the first cause, but they are never as violent as was the moment of impact. These vibrations, however, extend in all directions from the point of original excitation.

Vibration acts in the same manner, no matter what the medium may be in or through which it is expressed; and everything in this universe can be reduced basically to some rate and character of vibration.

So, in the instance of the human mind, emotional experiences there recorded follow a time sequence, but *with variations of intensity which correspond to the reaction brought about by the original happening.*

I, in a receptive state of mind, would be most apt to tune in on the *"peak emotional moment"* of any experience that Wilkins had recently registered in his consciousness. The proof has been abundantly recorded that this actually did happen, time and again. And there is every reason why, with telepathy accepted as a fact, that this *should* be so, since the most strongly vibrating thoughts should be the easiest to pick up.

But this is not always true. My mind, in the hypothetical case discussed above, might have contacted that "line of time," containing its series of mental pictures relating to the auto accident, at the point at which Wilkins stepped off the curb, and saw that he was going to be hit. The emotional impact I would sense in that event would not be so severe.

Yet, it might be sufficient to cause me to *hold* the impression, and to try to pursue it further. My inner mind, once focused upon it, might run along that "line of time," and might sense Wilkins hit by a car. But if that happened, I would *feel* the *increased emotional reaction* synchronized with the flashing picture or sensation of his being struck! And if I were able to "tune-in" on such a complete sequence, I might receive sufficient impressions of the whole happening to describe it accurately.

Now, let us reconsider an ESP card test: one's mind has no "line of time" *spread* to "tune-in" on. It must receive an *exact* impression of the symbol being held in the mind of another person. There is no emotionalized experience that can be linked with this symbol in order to project a thought wave or impulse with great and genuine intensity.

Furthermore, the receiver's mind is inhibited by the knowledge that he is to get an impression of one of five arbitrary symbols. It becomes exceedingly difficult, under those circumstances, to exclude the element of guess work. And I contend it is a far greater achievement for a "sensitive" to get consistently high scores under those test conditions than it was, say, for me to score as I found I was able to do in my experiments with Wilkins.

I believe that is true because, in a case in which one has no knowledge whatsoever of what may be received, one is able to relax more completely, and to develop greater confidence in the impressions as they come through.

In ESP work, the mind of the sitter is apt to be plagued by interloping conscious suggestions: "How do you know it's a circle? It might be a square, a cross, a triangle, or a wavy line . . ."

Once lost in the maze between conscious and subconscious, the receiver is no longer sure of his impressions, and his score may only run to "chance," or even below the normal expectation.

Absence of a strong, or of even an ordinary emotional factor is the greatest handicap. In trying to secure impressions of ESP cards from the mind of another, the receiver is unable to check a registration of the impression he is receiving against the *effect* any transmitted human experience has on his emotional system.

Such an effect, which is felt in the solar plexus region and accompanies the impression as it strikes the receiver's consciousness, carries tremendous conviction, and is one of his surest indications that he has really tuned-in on a thought or condition.

It is impossible for a sender, attempting to visualize mental pictures of a cross, a circle, or of another image, to get emotionally aroused about them as a means of increasing the energy discharge of the thoughts which he broadcasts.

Harnessing of the emotional system, as a power station behind thought projection, comes only through a vivid emotional experience which has been lived, or is being lived at the moment of sending.

I do not mean to disparage in any way the work being done by Dr. Rhine and his fellow scientists. But I do feel it necessary to emphasize the part one's emotions actually play in sending and receiving.

I am so convinced that emotions, in some mysterious way, are the "power generators" of extrasensory perception that I predict, were experiments of the kind I undertook with Wilkins tried out at universities, results easily as successful, and quite possibly more so, could be obtained.

Telepathy, as a faculty of mind, is much nearer the surface of consciousness, and much more operative in the life of man than he usually realizes, or lets himself admit.

# 4

# Unknown Powers of Mind

WHEN YOU HAVE LEARNED how to open up channels of extrasensory perception in your mind, you will find yourself receiving impressions, on occasion, which cannot be explained on the basis of telepathy alone—and sometimes, not at all.

For instance, Wilkins had only two accidents befall his plane in his entire time north, and my mind, in some incredible manner, brought in previsions of these happenings, days before they actually took place!

On January 27th, 1938, I recorded:

"Crankcase of plane comes to me suddenly . . . Did something go wrong with it, due to cold?"

On February 14th, I received by airmail from Wilkins this comment:

"Wonder if this was *forethought or 'preview knowledge,'* since it was not until *February 6th* that *we developed serious trouble in crank-case—main bearing of one engine ground to powder that day* . . . Must have been some trouble there since January 15th . . ."

January 15th was the date I had foreseen, on December 28th, 1937, as the time Wilkins would take off on a long search flight.

I had recorded, the evening of December 28th:

"*January 15th* comes to me as day you actually make take-off for north regions—*though you now hope to get off a few days earlier in month* . . . This again is *premonitory impression* as though thoughts jump, ahead of present moment, and *this future moment*, for a fleeting instant, becomes NOW . . ."

Later, on the evening of January 13th, the day after the moon was full, and the most ideal time, barring bad weather, for Wilkins to have made a search flight, I recorded:

". . . am again impressed with *15th of January* as your probable take-off date . . . that conditions won't permit attempt until then, despite your present *standing by* attitude . . ."

When Wilkins replied to a great number of my impressions on February 14th, he said of the one concerning his date of flight:

"QUITE EXTRAORDINARY that we DID make the flight, starting night of 14th (actually the 15th, your time) and returning on the 15th of January! We should have gotten off on the 13th, but radios on plane had broken down . . ."

The night of the 13th, I had even telepathically picked up the trouble, referred to, with his radio, which was reported and confirmed, as follows:

| SHERMAN | WILKINS |
|---|---|
| *Something of great importance mechanically has been found impractical and has had to be changed . . . I get this feeling strongly and see you supervising work in connection therewith . . .* | Radio |
| *You really need a new or different device which you don't have with you . . . something is ingeniously made at Point Barrow by your mechanic in association with your radio engineer . . .* | Also means of cleaning plane's windows of hoar frost—working on it intently. |

This illustrates how most strong thoughts, felt by Wilkins, could be tuned-in on.

But this ability of the mind to "see" what was going to happen in the future! How could it be accounted for?

The premonitory flashes had come unbidden amid the impressions of events which had already occurred. But they had brought with them a different sensation: I could usually tell whether a mental picture pertained to the past or the future.

# Unknown Powers of Mind

The feeling is difficult to put in words. It seemed to have something to do with my breath.

I would suddenly, and without warning while in a receptive state, feel my breath leave me—almost as though my consciousness was being sucked from my body. Simultaneously with this there would appear in my mind a vivid picture of a happening. I would view this scene, feeling myself to be momentarily in a state approaching *suspended animation*.

Once I had caught the impression, I found I could breathe again, and was conscious of having taken in a deep draught of air, settling back physically as though my consciousness had returned to my body. This all took only a comparatively few seconds of time, but the experience carried with it the sensation of my having been transported into another dimension in time. There just are not, quite obviously, the proper words for describing such an experience.

Your own mind has probably brought you similar knowledge of future events in the form of what you call "hunches," but not having been in the same sensitized state, you have consequently been unaware of the exact feelings incident to the phenomenon.

Even so, different individuals have told me that they have been "stopped in their tracks" by a sudden impression that has hit their consciousness, blotting everything else out, and leaving them with an inexplicable conviction that a certain thing was going to occur. The impression has been so strong that they have felt impelled to act upon it.

On the nights of March 7th and 8th, I had vivid impressions of a forthcoming accident, which did not take place until the morning of March 11th. Wilkins reported this narrow escape of himself and crew to the *New York Times*, by wireless, and it was published in the paper, March 12th.

It can be seen how my "prevision" impressions tally exactly, in every detail, with the news report of the accident.

| SHERMAN | WILKINS |
|---|---|
| *(Night of March 7th)* | *(Morning March 12th)* |
| | *New York Times* |
| | NEWS ACCOUNT |
| | WILKINS PLANE DAMAGED |
| | Ship Hits Snow Ridge Taxiing And Fuselage Is Torn |
| Was tail of plane slightly damaged in bumpy landing? *See some work*—in rear of plane . . . | AKLAVIK, N.W.T. March 11, (By Wireless) . . . |
| *(Night of March 8th)* | An odd freak of the weather today led to the *second* accident our airplane has sustained during our efforts to locate the missing Soviet airmen. |
| *Fleeting vision of your face—quite a* | |

| SHERMAN | WILKINS |
|---|---|
| *strained, intent impression as though concentrated upon flight activity in plane—seems as though* flight started and down at some point or turned back—plane motionless—something did not go as planned—snow or sleet-like weather some parts—seems to be pelting plane . . . | It was light and clear this morning until 6 o'clock, but shortly after we took off, a snow-laden squall, as black and sudden as a thundercloud, enveloped us, and fearing heavy snow and "icing up" on the machine, we were forced to land while we could. |
| Strange feeling in pit of stomach—or solar plexus, like I've gone through close scrape or acute experience—You concerned about something—*Slight break in clouds above*—but dark storm clouds low on horizon. "Carry supplies"—*these words come to me and feeling as though* supplies of some kind transported . . . | Pilot Herbert Hollick-Kenyon made a good safe landing with our heavy load of 1,200 gallons of gas and equipment, but in *taxiing back to our starting point*, we struck a solid sharp ridge of snow and the *tail-skid* was torn from the fuselage. |
| | Engineers A. T. L. Dyne and S. A. Cheeseman, whipping back and forth between the machine and our main base in a home-made propeller-driven sled, are quickly effecting repairs, and, by working throughout the night, expect to have the machine completely repaired by tomorrow. |

So definite were these impressions of the accident, four days in advance of the happening, that I had even gone on to record on the night of March 8th:

> I see you (Wilkins) beside the plane, looking up and around. There appears to be a *ridge* or slope beyond, of snow and ice—in ribbon-like outlines—There is activity about a camp or tent not far distant—some men moving in and about . . .

Here was described what must have been Wilkins' actions when he climbed from the plane, and examined the damage done to the tail, in landing, against the *ridge* he reported in his news story, and which I had mentioned in my impressions three days before!

There was also the reference by me to the activity of men in the vicinity of the plane—members of his crew whom he told of carrying supplies from camp to plane, and my recorded impressions which I put in quotes, "Carry supplies."

Wilkins further substantiated, when he returned some weeks later, that I had described exactly the feelings he underwent as the plane, fully loaded, was coming down for its hazardous landing.

With the evidence of this ability, at unexpected moments, to pick up previsons of coming events, I was led to ask myself: "What *is* Time?"

Were the happenings I had foreseen destined to occur? Could nothing have intervened to prevent them? Had all of their causative forces been set in motion, and had I but "tuned-in" on these events, taking shape in another dimension?

We know the future exists for us in some form—that a definite series of things are going to keep on happening to us from this moment on—up to death—and possibly beyond.

Certainly a chain of events, infinite in character, covering millions and billions of years—inconceivable in number—led up to the time of our individual birth, an event, which no prophet would have dared predict.

Had my mind, in its sensitized condition, been reaching out and making occasional fleeting contact with realms of intelligence and action to which we, as humans, are intangibly, but nevertheless actually, related?

Does this unseen universe which lies around and about us contain the answer to all our riddles of human personality and identity, once we learn how to delve into the almost frightening and certainly awe-inspiring depths of our own selves? These questions arose in me as I opened my mind, allowing my inner consciousness to seek impressions from without.

And there has come to me now, after years of experimentation, what I believe to be a true explanation of certain phenomena of mind.

I am convinced that we are *constantly creating* our own *future* by the *nature* and *character* of our *thought*—projecting the *inner* self *ahead* of our conscious outer self, setting it to work *paving the way* for *events* which are prepared for us along the banks of the stream we call *time* . . . And we are carried *to* these events, slowly or quickly, in *exact accordance* with the *intensity* of our thoughts *concentrated* in *that direction*.

Our *desires, ambitions, fears* or *worries*—all our *emotional reactions* to *external experience*—set up *vibrations* inside ourselves which commence to *attract* to us *conditions, circumstances,* and *people* after the *nature* and *character* of our thoughts.

How our subconscious mind has the power to bring all these outside circumstances into line is only one of the countless miracles—but my observation is that it does—it is *constantly* doing it. *Myriad thoughts* are interchangeably at work among individuals who are later destined to meet—the *law of attraction* relating them on a plane of *mental activity* of which we are now but vaguely conscious, and which exists in a *"potential time dimension."*

*Certain acts* and *thoughts* of ours are *attracting* to each passing present moment experiences built up for us in this *"potential time dimension,"* so that these experiences—once drawn to us—are continually becoming *realities,* having lain in wait in answer to the *bidding* of our *strong desires, ambitions,* or *fears* to transform themselves from a *"future possibility"* into a *"present fact."*

Under these conditions it can be seen how I, or any developed sensitive, might "tune-in" on events of which the inner consciousness of an individual is aware—events that are shaping themselves to come to pass in "future time."

Much remains to be explained, of course; and no scientist has as yet been able to define the processes which take place in the inner mind in order that the "future" may be brought to us in orderly, progressive sequence in the form of each succeeding "present moment."

And, furthermore, it is a fact worth noting that each individual's future is distinctly different from that of any other—however closely he may be related. His reactions are different; the specific things that happen to him, physically and mentally, over a period of time, are not exactly the same, or in the same degree.

All of which would indicate that the mind of each person, in some mysterious way, creates and attracts the conditions and events with which the physical self is to become associated on this earthly plane, in future moments of time.

If this be true, then there is every reason for our undertaking a study of these inner powers of mind to the end that our discovery of the laws behind their operation may eventually bring us mental and physical freedom from the wrong application of these very laws in our everyday life.

# 5

# Keeping One's Mental and Physical Balance

BY DEVELOPING THE FACULTY of telepathy, you are increasing the sensitivity of your nerve reaction to conditions both within and without your body.

Unless you possess a good, well-balanced nervous system, it is well not to attempt intensive experimentation in extrasensory perception.

I paid dearly, on the physical plane, for my strenuous schedule with Wilkins.

Had I not been forced, by economics, to continue my writing activities during the five months' test period, and had I been able, instead, to devote my mind to the experiments freed of each day's ordinary mental and physical strain, I should not have overdone.

As it was, making this highly concentrated effort to receive thought impressions from Wilkins three nights a week, preparing the necessary copies and records of these notations, and getting them in the mail at once to the proper witnesses, took not only time into the small hours of the morning, but also a severe toll of nervous energy.

Despite a developed ability to relax my physical body, and to gain a new supply of energy with a few hours' rest, I was making almost superhuman demands upon my mental and physical machine.

Then, too, as the experiments progressed, I could not control a growing feeling of responsibility which put pressure upon me.

It was becoming more and more apparent that I was scoring a high percentage of "telepathic hits" which only increased my desire to maintain or improve this average, and to guard against interference by my conscious mind, or intrusions from my faculty of imagination.

As a writer, I knew the "feeling in consciousness" engendered by the imagination, and I could usually recognize it whenever the imagination became awakened and tried to "color" a telepathic impression for me. .

But inducing and maintaining a sensitized state of mind, capable of making this distinction in consciousness, was a strain in itself, since it burned up nerve energy.

I had to *"still* my own emotions" during my periods of receptivity, holding the nerve centers of my body so that impressions from the mind of Wilkins might register upon them, once we made subconscious contact with each other.

After I had received the "previsions" of accidents which I felt were going to happen to Wilkins' plane, the human side of me was affected. I was personally fond of Wilkins as a friend, and my emotions tried to tell me that he might meet with some serious physical injury, or even death.

This brought on a real battle with myself each appointed time. I found instilled in me an emotional apprehension that I might "tune-in," at any sitting, and receive impressions of a possible fatal crash. This apprehension had been brought about through the intensity of the conviction in my inner mind that what I had foreseen would really come to pass. I had no doubt of it. I could not determine the exact time when my premonitory impressions would materialize, but I felt they were imminent in the "very atmosphere"—a fact which proved to be true.

The effect on me was almost nerve shattering. On each night of a sitting, it required a strong effort of will to put my own emotions out of the equation, and to "blank my consciousness" so that I might receive, "uncolored" by my own fears, impressions of what was actually happening.

Obviously, to have permitted these apprehensions to have dominated my consciousness, would have prevented the reception of any genuine impressions at all.

I experienced a tight, tense sensation in my solar plexus region, with occasional shortness of breath, and nervous indigestion, followed by stomach cramps. This condition, despite all efforts to throw it off, became chronic.

I walked the floor many nights with severe stomach pain and "gas pressure," which defied relief. It seemed that my entire stomach and central nerve area was in a knot.

I could not and would not abandon the experiments, and simply made a note of my physical reactions for the records, while keeping on with the tests.

I consulted my physician during this period, and he told me that my condition could not be alleviated until I was able to get out from under the mental load I was carrying. I was "highly keyed up," ordinary happenings—such as a slammed door or the sudden backfire of a car in the street—striking me like a knife in the solar plexus. My entire nervous system was greatly "oversensitized," and reacted involuntarily to conditions which normally scarcely influenced it at all. Yet all that I was experiencing was of profound value to me, and of marked significance.

## Keeping One's Mental and Physical Balance 141

When the tests were concluded, the relief from strain was almost too much to endure. I had to let myself down easily and gradually like a horse that has run a furious race, and must now be walked about and cooled off by degrees before being returned to the stable.

I continued, at times, to receive unsought impressions from the minds of others, when I fixed my attention upon them.

And, for some weeks afterward, at the appointed hour of eleven-thirty on Monday, Tuesday, and Thursday nights, no matter what I might be doing consciously, my mind would jump to Wilkins!

And when I would be about to drop off to sleep at night, putting my mind in a passive state, and thinking about certain persons or things, there came to me, unbidden, flashing impressions. Some of them were so vivid and compelling that I got up and wrote them down, making them a part of my private records for later study and research.

All of these facts serve to indicate that when you develop additional faculties of mind, and set them methodically in motion, you will have difficulty in subduing them—once the intensive work is over.

We know, of course, in the physical world that athletes do not quit strenuous competition in sports suddenly. They take a few seasons to taper off, experience having told them that this process is necessary to permit the system to adjust itself, without resultant damage to the body.

Apparently, to a certain degree, the same is true in the operation of mind.

With me, however, considerable physical damage had already been done. I had developed ulcers of the stomach, due to this prolonged nervous strain, which, in turn, produced several hemorrhages. My life, itself, was threatened, before I was able to regain proper control of my nervous system.

This knowledge that our thought processes are so definitely tied up with our emotions should cause us to consider the continuous effect upon our lives—physically and mentally—of wrong emotional reactions. Many of us have had nervous breakdowns, or have been on the verge of them, through letting our emotions get out of control. A well-ordered body is the product of a disciplined mind. We must have both, if we are to realize our fullest degree of mental and physical efficiency.

What I have gone through on the physical side has demonstrated certain truths to me of practical value to every individual. Living, as we are, in these uncertain economic times, many of us are under great daily nerve strain, and we need to know how to gain at least temporary mental and physical relief, which can be done by setting aside a time each night, with the day's work finished, for quiet communion with our inner selves.

This is one of the rarest experiences possible to humans and yet few are availing themselves of it. I look forward each night to this moment as an opportunity for withdrawing from this outer world, of getting outside myself, and of regaining a true perspective—an emotional and mental balance. It is so easy to get "off-center" about things under the stress and confusion of everyday life.

Most of us are carrying, at all times, our concern over the physical health of our loved ones, as well as pressing financial burdens. Our bodies, reflecting the attitude of our minds, become tense and highly emotionalized.

Under such conditions, fears and worries are intensified, and we cannot help visualizing the possibility of dire things happening. In such a sensitized state, we are actually attracting the very things we *don't* want to happen, through fear that they *will!*

For this reason, it is imperative that we keep this evening appointment with ourselves—to check the influence of wrong emotional reactions which we have permitted to appear during the day.

The first step is to prepare for retirement, and stretch out upon the bed, devoid of clothes. In fact, it would be better if we wore no night clothes at all, at any time.

Give your body this new and complete sense of freedom during the night hours. It has been confined in clothing all day long. And if you have felt hemmed in by oppressive circumstances mentally, your clothing has had the same effect upon you physically.

Your whole desire, as you approach this moment of meditation, is to "shake off" the unpleasant and disturbing emotional hang-overs of the day. Now, lying *relaxed* by the method already described in an earlier chapter, you are ready to clear your body and consciousness of the conditions you have "taken-on," which are exerting a destructive influence.

You have thought, before you tried this method, that you just could not stand it, if you had to go on much longer this way, fighting your problems by day, and wrestling with them all night. You would give anything if you could get some release from it all, and enjoy a good night's sleep to prepare you to face the morrow, filled with new nerve energy and a fresh outlook.

What happens when you are asleep is one of the deep mysteries. But we have definite knowledge that the subconscious part of us never sleeps. Our job is to give that subconsciousness a set of the right kind of mental pictures to go to work on, each night, while we retire from the battle, in the calm, confident faith that reinforcements are being brought up, and the weak spots in our armor repaired and replenished by the creative power within.

To control this action of our subconscious, we must, in this self-induced meditative state, *re-live* the experiences of the day in our *mind's eye*. Call to consciousness again the mental pictures now stored in memory of the things that have happened to you. See yourself undergoing these experiences as though you are a spectator on the sidelines. Check your memory of how you have reacted to these experiences. Do you think you could have done better? Did you make some mistakes? Did you lose your temper, incur the ill-will of someone, indulge in hates, or develop a fear of certain conditions?

All these wrong reactions must be recognized impersonally, and weeded out of consciousness, or else the creative power of your inner mind will accept

# Keeping One's Mental and Physical Balance

them as conditions you want in life, and will attract similar experiences to you in the future.

Whenever you come upon a recollection of some incident that clearly indicates you were in the wrong, *immediately visualize* a mental picture of how you *now* know you *should* have reacted to this experience, so that your new mental picture can replace the destructive one in consciousness.

The fears and worries you feel at the moment were born of just such unhappy experiences, and can only be eliminated by going back to the source in consciousness, and recognizing the wrongness of your own reactions. The instant you have done this, and corrected these fear pictures, they lose their hold over you.

You must realize, each night, that the thoughts you think represent the *only* power you possess. Then why give power to the wrong kind of thoughts? Do you think your loved ones are being helped by your concern over them? Each person has his life to live; he must face his situations basically alone, and we cannot help him or ourselves by worrying.

Did it do me any good to be apprehensive over what might befall Sir Hubert Wilkins in the Arctic? Whatever was going to happen, he had to meet his own experiences in the way that he was prepared to meet them.

I was not experienced enough, in this pioneering work, to eliminate entirely my feelings of anxiety under the highly sensitized conditions of body and mind which existed.

One day I hope to have attained such development that I can be impervious to what may happen to others, in my inner mind. By so doing, I free them and free myself to face life to the best of our capacities.

In the ordinary rounds of life, I have learned how to exercise emotional control—how to free my mind of hates and prejudices, of the usual fears and worries.

And what I have been able to demonstrate, you can also do, with a little conscientious practice. What we call extrasensory perception is operating in us all the time. We are constantly "tuning-in" and out of wrong and right conditions in our minds. Our emotional reactions to external experience are determining our future.

It behooves us, then, to acquire *conscious control* of these emotional reactions in order to keep the wrong kind of pictures from reaching our subconscious, and thus attracting more and more destructive happenings to us. By developing such a technique, we are making practical use of these higher powers of mind, and are benefitting ourselves immeasurably!

# 6

# Practical Use of Telepathy

WE ARE IN THE beginning stage in respect to our knowledge of the higher powers of mind. Our position may be likened to that of Marconi when he sent the first feeble wireless wave winging its way across the Atlantic Ocean to be picked up on a crude receiving set, thus demonstrating that a new dimension of communication was open to man.

What is being revealed now through the research of universities and medical laboratories, as well as through the efforts of independent scientists and investigators, is laying the groundwork for a new and great science of the mind that will scrap many age-old theories of the nature and character of consciousness and of brain operation.

One is no longer met with widespread skepticism and scoffing when the subject of mental telepathy is introduced.

Radio, and its power to operate through the invisible medium of the air—the fact that sounds are all about us, yes, and electrically vibrating images of actual persons and scenes, too—has done much to do away with public incredulity, and to cause humans to accept that "anything is possible today!" This is a most promising state of mind for humanity, since such a receptive attitude opens the door to new wonders beyond, and invites their materialization in this, our physical world.

Man must first believe that "all things are possible" to him, within the range of his human capabilities, before great achievement can be his. We suffer only the limitations that we put upon ourselves through our own limited concepts. And so, repeated ignorant or stubborn denial of the existence of certain powers does not keep them from existing—except for *us!*

Recognition of the laws behind the operation of extrasensory perception,

# Practical Use of Telepathy

and knowledge of how to bring these higher powers of mind under conscious directional control, are going to be the next great step in man's upward progress.

True knowledge of mind, of his own inner self—with man's faith supported by the findings of science—will do more eventually to bring about the centuries-old dream of universal brotherhood than any other intelligent force.

We cannot understand our fellow man, nor he us, until we have come closer to solving the mystery of our own selves, our relation to each other, and to this universe.

Such development is still far in the future, but, for many enlightened humans, enough is now known to give them practical utilization of these powers of mind through earnest exercise and application.

Telepathy, as I have shown, is just one phase of an apparently unlimited and still uncharted field of extrasensory perception. Much experimentation and investigation remains to be done before we can speak authoritatively of the manner in which certain information is made available to our minds through one or more of a number of extrasensory perceptive faculties.

But we do not have to wait until science can explain what happens behind the scenes in our mind's operation in order to get evidential results. Science is still unable to tell us basically what electricity is—and yet it is constantly expanding its constructive use of this force or energy.

So it is—and will be—in the development and unfoldment of mind!

Mark Twain, in the year 1906, had need of an article he had written for the *Christian Union*, published in 1885. He searched his files, but could not find a copy.

Mark's desire for the article was great. He wished to base a new series of articles upon it. He *had* to get a clipping of it somewhere. But there was not even a copy to be secured from the office of the publication. Mark went exhaustively through all his papers. No result.

Some days later America's greatest humorist had to be in New York City on business. He was walking down Fifth Avenue, and was stopped at Forty-second Street by traffic.

While he was standing, waiting, a stranger rushed up, and addressed him.

"Mr. Clemens!" said this man. "You don't know me—but here is something you may wish to have!"

With this, the stranger pressed some clippings into Mark's hands.

"I've been saving these for more than twenty years," he went on. "But *this morning*, it occurred to me to send them to you. I was going to mail them from my office—but now I'll give them to you."

Mark thanked the man, who disappeared in the crowd of passersby. Then he looked down at the clippings.

They were the very ones he wanted—from the *Christian Union* of 1885!

A coincidence? The chances are millions to one against it.

From what we now know of the operation of the mind, it is more logical to assume that Mark Twain, having instructed his subconscious, through his strong desire, to secure for him the clippings of his article from the *Christian Union* of 1885, had activated his extrasensory perceptive powers.

In some way which we can demonstrate but not explain, there was "broadcast" the mental pictures of these clippings, and of Mark's need for them. Since "like attracts like," this broadcast reached the consciousness of an individual who possessed a set of the desired clippings.

He was impelled to think of them, and the impression he received was so strong that he was also given the feeling Mark Twain might like to have them.

The stranger could not distinguish the difference between impulses reaching his conscious mind from outside and those coming from his own subconscious, and naturally regarded this sudden thought and urge as his own idea.

In his mind was the mental picture Mark Twain had unconsciously planted there—the picture of Mark's securing a copy of his article from the *Christian Union*.

Mark's earnest desire had painted that picture; and the intensity of his emotional yearning for the clippings had generated the power that had sent it out, seeking its affinity in the consciousness of someone else who possessed what Mark wanted.

This man felt impelled to reproduce this picture in actual life, once he had received it, by getting his clippings of the *Christian Union* article, and sending them on to Twain.

*But*—as if this was not remarkable enough—witness what else happened!

Once the subconscious minds of Mark and this stranger had made contact, the "circuit" had apparently remained "closed," causing the physical movements of both men to be *so synchronized in time and space* as to have brought them to the *same* point on Fifth Avenue at the *same* moment, thus enabling Mark personally to receive the desired clippings!

Once the mental picture was *materialized* through this action, the magnetic attraction between the two minds ceased.

The stranger, having discharged his impulse, was freed from it, disappearing in the crowd without even identifying himself.

And Mark was left with the clippings he had visualized himself receiving, his own mind relieved of the intensive desire which had existed there, which had kept the picture vitalized until its fiulfillment in this "outer world."

"Coincidences" of a similar nature are occurring to many people every day, to people who unconsciously operate the law of the subconscious mind correctly, and are obtaining outstanding results. But these happenings are actually the mathematically realized effects, growing out of causes they have set up in their own consciousness.

In the fall of 1936, I was staying at the City Club of New York, on West Forty-fourth Street, for several months. During this time I had a need arise to

## Practical Use of Telepathy

get in touch with a friend, a former landlord, owner of an uptown apartment house, whom I had not seen in years.

I phoned the apartment house only to learn that James J. Bradley had sold it several years before, and no one knew where he could be reached. I looked in the phone book, but his name was not listed. There was apparently no ready way to locate him.

I then sat down and visualized myself meeting Mr. Bradley. I saw a confident, expectant picture in my mind's eye of my "running into him" somewhere. Once I was sure my subconscious mind had received this picture and would act upon it, I "forgot about it" and went about my business.

Each day, as Bradley came to mind, I repeated the picture, "seeing the two of us meeting, being drawn together." I wasn't overanxious to encounter him, or apprehensive that it would not happen. To me, in my inner mind, it was an accomplished fact.

It had been my habit for weeks to arise and leave the City Club for breakfast around eight o' clock in the morning.

One particular morning I awakened, having no disposition to get up and out on time. I felt impelled to write several letters, and to attend to some odds and ends. I was surprised to find it close to nine-thirty when I finally started out for breakfast.

Now, suddenly, I found myself in a hurry. I could not leave the City Club quickly enough, and almost ran out the front door.

As I did so, directly across the street from me, a man was passing. He looked at me, and I looked at him.

"Why, hello, Sherman!" he called, and came over toward me.

It was James J. Bradley!

I was more thrilled than amazed, for incidents of this kind had been happening to me for some time.

Note how my subconscious mind had apparently made contact with Bradley's subconscious, without my conscious knowledge, and had remained aware of his physical movements, either giving him the impulse or knowing that he was going to pass the City Club at this moment in time, and causing me to *delay* going out to breakfast for an hour and a half, so that *my* physical movements would be synchronized with Bradley's as he went by the door!

Another way in which your subconscious mind often produces what you picture for it, is for it to cause you to make inquiries concerning a certain person or a thing you want from someone you may consciously think knows nothing whatsoever about the person or thing, and could be of no help, only to find that he is the *very person* to provide the connecting link between yourself and that which you are seeking!

The evidence is inescapable, but how the inner mind does it cannot be technically explained. It is a phenomenon, however, that can be reproduced at will, once the law of your subconscious mind is understood and properly operated.

Charles T. Lark, a well-known New York attorney, sold a gorilla some years ago for a client to the Ringling Brothers–Barnum & Bailey circus. The gorilla was highly advertised as "Gargantua."

Mr. Lark was not a theatrical lawyer. His business was ordinarily the handling of wills and estates. Because this transaction was so unusual, it made an indelible impression on his mind.

Recently a wealthy woman from Cuba, in New York to attend to personal matters, was referred to Mr. Lark by the Chase National Bank, when she asked to be sent to "a good lawyer."

"I suppose you've never handled a case like this," she said, when she entered Lark's office.

Then she revealed that she was the owner of a nine-year old female gorilla which she and her husband had captured as a week-old baby on a hunting expedition in Africa. The gorilla's parents had charged them, and had been shot. The life of the baby had been saved through turning the young gorilla over to a native woman to nurse.

But now the gorilla, having been brought to Cuba and raised as a pet, was growing unruly. Its owner wondered if Mr. Lark could help her dispose of it to the Ringling Brothers–Barnum & Bailey circus as a possible mate for Gargantua.

"I should think I can, if anyone can," said Mr. Lark, "since *I* was the lawyer who negotiated the *other gorilla deal* with these circus people!"

Asked why she came to him of all the thousands of lawyers to whom she might have gone, the amazed woman said she did not know. She had wanted to sell this gorilla to the circus, and hadn't known how to go about it, but, on the way to New York, it suddenly occurred to her that she would ask her bank to recommend a lawyer.

She made this request without disclosing the nature of her business, and the banking official sent her to Mr. Lark without knowing of his previous "gorilla experience."

Can any honest investigator put these happenings down as "mere coincidence"?

These faculties of extrasensory perception are helping us weave the pattern of our lives in answer to the nature of our desires, attracting to us *like* conditions, circumstances, and people in *exact* accordance with the *kind* of mental pictures we implant in consciousness!

Thomas Edison, asked how he invented, said he had trained himself to think in terms of mental pictures. He always *saw* a "picture" in mind of what he wanted to invent—before he knew how to do it. Edison assimilated all the known facts on a subject with his conscious mind—then turned them over to his subconscious, and went to sleep on a cot which he kept in his laboratory.

Then he followed the "hunches" his subconscious gave him when he awakened, and often returned to consciousness with the entire solution or idea for many of his inventions. Edison had learned to *consciously* control and direct the action of his subconscious.

And, like Edison, you must have a definite design or picture in mind in order to get a definite result in your external life!

I do not pretend to know the answers to much of what occurs in extrasensory preception. But I do know that these powers of mind are not freakish endowments of a perverse Nature. We *all* possess the same faculties capable of development.

Man's mind has brought us all the great inventions, the great music, the great works of art, the civilization of which we are a part today.

And Sir Hubert Wilkins and I are convinced, through our exploration of human consciousness, that the new, *real* history of the world will not be written on man's conquest of man, but on man's conquest of his own mind!

## Part Three

# Authenticated Documentary Record of the Wilkins-Sherman Experiments in Long-Distance Telepathy

*Part Three comprises a day to day chronicle, taken word for word from the original record, of all the "telepathic hits" scored by Harold Sherman in his five months' period of scientifically witnessed experiments in extrasensory perception.*

*Included also, are the commentary made by Sherman at the time of each sitting, which he felt might throw some light on his physical and mental condition at the moment as well as upon any other factors having a bearing on the nature of the results obtained.*

*This record provides an interesting and exciting study, and possesses unusual significance for the serious investigator of mental phenomena.*

State of New York )
County of New York ) ss:
City of New York )

When Harold Sherman informed me that he was attempting to get long-distance telepathic impressions from Sir Hubert Wilkins in the Arctic, I told him I would be glad to receive and file his daily communications regarding his impressions. He proceeded methodically to record and mail to me the impressions received. The postmarks showed that he was giving me an up-to-date record. I have all these communications in my file with their original postmarked envelopes. I am not able to evaluate the series of impressions as a whole and this statement is to be construed only as testimony that Sherman did send me promptly the dated impressions which he received.

Subscribed and sworn to before me this
11th day of June, 1938

*Gardner Murphy*

*Nancy D Baines*

NOTARY PUBLIC, New York County
N. Y. Co. Clk. No. 20
New York County Register No. 0B118
Commission expires March 30, 1940

TESTIMONY OF SAMUEL EMERY,
LAY WITNESS RESIDENT
CITY CLUB OF NEW YORK.

This is to certify that I, Samuel Emery, resident of the City Club of New York, 55 West 44th Street, New York City, being a friend of both Sir Hubert Wilkins and Harold Sherman, was asked by them to be a lay witness of their telepathic experiments and that I received from Harold Sherman complete typewritten copies of his impressions the day following their reception. I was able subsequently to check them as confirmation was eventually received from Sir Hubert Wilkins, and I can testify to the honesty and sincerity of Harold Sherman, the receiver, who could not possibly have access to the intimate knowledge of Sir Hubert Wilkins' many personal and expedition activities in the far north except through the agency of Extra Sensory Perception.

(Signed) Samuel Emery.

*Samuel Emery*

May 9, 1938

I herewith attest to the genuineness of Mr. Samuel Emery's signature.

*John J. Cassidy*
Chief Clerk - City Club of N.Y.

TESTIMONY OF DR. HENRY S. W. HARDWICKE,
RESEARCH OFFICER FOR THE
PSYCHIC RESEARCH SOCIETY OF NEW YORK

This is to certify that I, Dr. Henry S.W. Hardwicke, have been in weekly contact with Harold Sherman throughout the period of the telepathic tests he conducted with Sir Hubert Wilkins; that I have known of many of Mr. Sherman's recorded impressions shortly after they were received by him and weeks before Sir Hubert Wilkins could be reached for the purpose of determining and verifying the experiences he had undergone as described by Sherman on those dates. There can be no question of the authenticity of this telepathic phenomena. The distance alone, between sender and receiver, of over two thousand miles - and the fact that most of Sherman's impressions pertained to Wilkins' activities on the very day of their reception, should answer any reasonable skepticism - it being perfectly obvious that it would be humanly impossible for any person to accurately record the experiences of another, at such a distance with the time element so closely synchronized - and accomplish such a feat, consistently, week in and week out for a period of six months - without the exercise of a telepathic faculty.

(Signed) Dr. Henry S.W. Hardwicke

*Henry S. W. Hardwicke M.D.*

NOTARY PUBLIC, Bronx County
Bronx Co. Clk. No    Reg. No.
N.Y. Co. Clk's No. 410. Reg. No.
Commission Expires March 30, 1940

TESTIMONY OF DR. A. E. STRATH-GORDON.

This is to certify that I, Dr. A.E. Strath-Gordon, have had occasion to check Harold Sherman's recorded impressions during the six months' telepathic experiments conducted with Sir Hubert Wilkins in the far north.

On two different nights, that of February 17th, 1938, and March 24th, 1938, I sat in Mr. Sherman's darkened study at 380 Riverside Drive, New York City, and witnessed his actual recording of impressions as he received them from Sir Hubert Wilkins.

Mr. Sherman wrote rapidly and filled a number of pages of his notebook. My presence in the room did not seem to inhibit his ability to place himself in a "receptive mental state", because most of his telepathic impressions, recorded on these two nights, were subsequently proven to be what I would term, "photographically" accurate.

In my many years of study and research all over the world, in the field of mental and psychic phenomena, I have never observed such continued clarity and exactness of telepathic vision as that demonstrated by Harold Sherman. To witness his receiving and recording thoughts or thought forms, is to give one the feeling that Mr. Sherman is taking what amounts to dictation from some invisible intelligence.

The test conditions under which these Wilkins-Sherman Telepathic Experiments were conducted, were ideal. Sir Hubert Wilkins was, for the greater part of his six months' expedition in the far north, beyond the regular reach of airmail and short wave radio.

Thus, any possible day to day channel of communication, other than that of Extra Sensory Perception, was automatically ruled out.

Before me, *Henry C. Borger* On this day appeared a Notary Public in and for the *Allendale* County of *Bergen* In the State of New Jersey, the deponent, A.E. Strath-Gordon, who states that the above, to the best of his knowledge and belief, is true and correct, Subscribed and sworn to on this fourteenth day of June, 1938.

*Henry C. Borger*
*Notary Public N.J.*

*A. E. Strath-Gordon*

My commission expires *Apr 17/40*

### TESTIMONY OF REGINALD IVERSEN, NEW YORK TIMES RADIO OPERATOR

This is to certify that I, Reginald Iversen, Radio Operator for the New York Times, was in contact with Harold Sherman off and on during the period of his telepathic tests with Sir Hubert Wilkins. It had been thought that some of Sherman's impressions could be checked by short wave with Wilkins and thus expedite the report on the tests, but magnetic and sun spot conditions were so bad during this entire time that I was unable to communicate with Sir Hubert Wilkins except on a comparatively few occasions.

One Monday evening, February 21st, 1938, my wife and I visited Harold Sherman in his home and were present in his study at 380 Riverside Drive, New York City, when he was receiving impressions from Sir Hubert Wilkins and, at that time, Mr. Sherman recorded the impression that Sir Hubert Wilkins was trying to get some messages through to me by short wave radio. I was dubious that this was so because Wilkins knew that the next two days, Tuesday and Wednesday, were my regular days off duty at the Times, and he rarely tried to contact me when he was certain that I was not on the job. But I learned the following morning that these messages had been received the night before by our night operator at the Times, who had tried to reach me by phone, and that the messages contained additional information which Harold Sherman had also telepathically received and recorded in my presence.

At no time during this period of six months did Harold Sherman ever seek such information as I might have known concerning Sir Hubert Wilkins and his activities in the far north. In fact, despite my skepticism, as it turned out, Sherman actually had a more accurate telepathic knowledge of what was happening to Wilkins in his search for the lost Russian fliers than I was able to gain in my ineffective attempts to keep in touch by short wave radio.

(Signed) Reginald Iversen

Sworn to before me this 9th day of June 1938

CHRONOLOGICAL REPORT OF NEW YORK TIMES SHORT WAVE RADIO
COMMUNICATION WITH SOVIET SEARCH FLIGHT EXPEDITION
HEADED BY SIR HUBERT WILKINS

Due to the emergency conditions prevailing, Sir Hubert Wilkins left on his second search flight for the lost Russian fliers, with much of his equipment having to be arranged for and shipped to Canada after him. It was one of my duties to assist him in lining up this equipment, acting upon instructions from Wilkins.

Because of unprecedentedly bad communication conditions extending over the entire northern hemisphere - brought about by magnetic and sunspot disturbances - our intended schedule of short wave with Wilkins was almost completely disrupted.

There were long periods of time when contact with Wilkins at Aklavik or Point Barrow was made, but no traffic could be handled. On many occasions, the signals did not come through at the appointed times, and when they did, they were so weak that we could pick up no messages. But the dates when we did make successful contact with Wilkins, and received press dispatches from him, were as follows:

```
December  2, 1937    March  2, 1938
January  11, 1938    March  4, 1938
January  24, 1938    March  7, 1938
January  27, 1938    March 10, 1938
February  3, 1938    March 11, 1938
February 17, 1938    March 14, 1938
           March 15, 1938
```

The balance of the dates on our log from October 25th, 1937, to March 18th, 1938, when Wilkins' radio station at Aklavik was dismantled, show ineffective attempts to contact Wilkins, with signals too weak for communication, or signals "unheard" or "unreadable," or contact made for testing and little or no "traffic" handled.

It should be borne in mind, however, that the news contained in these press dispatches had, in most instances, happened some days before, and Harold Sherman, recording his telepathic impressions three nights a week, had already noted whatever he had been able telepathically to pick up concerning these same events, the copies of his impressions having previously been mailed and in the hands of Doctor Gardner Murphy of Columbia University and Samuel Emery of the City Club of New York.

Even so, I never notified Mr. Sherman on these few occasions, or at any other time, when communication of any sort took place, Mr. Sherman's first knowledge of my having received a press dispatch came when he saw it published in the New York Times.

Signed _R. J. Iversen_
Chief Opr. N.Y. Times Radio Station
June 12, 1938

## Test 1

### October 25, 1937.   11:30–12:00 P.M.

*I sat in my study at 380 Riverside Drive, New York City, notebook before me on desk, lights out.*

*I awaited impressions, either in the form of mental images or strong feelings or flashes which could be translated as having a certain meaning.*

*The following is an exact transcription of what seemed to "come through" just as put down by me at the time and contained in my notebook:*

| SHERMAN | WILKINS |
|---|---|
| *You late in starting* *Break away from others* *Trip satisfactory so far* | Learn today that I must leave Winnipeg and go to Montreal and Ottawa, via New York. The arrangements to be made there may delay our start. |
| *Equipment not ready* | Wireless equipment is not ready. |
| *Latitude 52—* *Longitude 99—* | Our Longitude about 99—Lat. 55. |
| *Delay impression* *One man short for expedition* *Delay impression again* | Another wireless man will have to be hired if I go there with the plan about which I shall visit Montreal. Two men suggested by telegraph. |
| *On to Aklavik Wednesday* *Equipment not complete* *Delay* | On to *New York* Wednesday to complete equipment—which means delay. |
| *You in company heavy-set man or he nearby—impression, as you would say, "Wilkins signing off"—writing this in darkness—do not know exact time* | At dinner given by Manitoba League of Aviators, two heavy-set men sat on either side. My address was given on the radio—the first half hour to all of Canada—and the second half hour to |

| SHERMAN | WILKINS |
|---|---|
| | only Winnipeg local stations. The "Continental" man did actually say, "Wilkins now signing off." |
| *Seem to see you get up and leave a room—and go out and join three men—yes—there are more than three—quite a group.* | After dinner I invited three men to my room. Shortly after, Kenyon came in with about fifteen men. |
| *You have had a hard time keeping this appointment—your mind full of plans and delay—something mechanical hasn't arrived or isn't satisfactory.* | Did not want to keep this dinner appointment. Tried to get out of it to go to Ottawa on Tuesday (today) instead of tomorrow. Thought that if I did not get to Montreal or Ottawa as soon as possible, our plans would be delayed. |

## Test 2

### October 26, 1937.   11:30–12:00 P.M.

*Due to the illness of Charles E. Whitmore, my very dear friend, I was compelled to stay at the City Club, 55 West 44th Street, New York City, all night.*

*Mr. Whitmore was at the New York Hospital, but his brother, Howard, was expected in from Boston on an early morning train which I was to meet with Samuel Emery, my witness in these experiments.*

*I occupied a room across the hall from his room in the club, number 709.*

*At 11:30 P.M. despite the fact that my mind had been filled with other matters, I retired to the room, and cleared my consciousness as best I could in the attempt to receive telepathic impressions.*

*I turned out the light after seating myself at a table, with yellow sheets of copy paper in front of me, on which to record whatever thoughts might come.*

*The following is an exact transcription of my impressions:*

| SHERMAN | WILKINS |
|---|---|
| *Same location as last night You sit with your eyes closed and will thoughts to me . . .* | Was on plane on way to New York. Was in air between Minneapolis and Chicago at appointed time. Thought I might get a chance tonight to concentrate but an |

| SHERMAN | WILKINS |
|---|---|
| | elderly woman sitting next to me insisted on talking, talking. |
| *Weather colder—has been mild—work all day on skis . . .* | The men were working on the skis. |
| *K comes strongly . . .* | Kenyon, was, of course, in Winnipeg. |
| *You seem to be in building with steps leading up to door . . .* | Fort Garry Hotel (in Winnipeg) has wide steps leading up to door. These steps remain in my mind as the most vivid impression when thinking of the hotel. |
| *K testing skis—Are they strong enough to support plane? This important question.* | This is a question—Skis designed for 10,000 lbs. Our load will be about 13,000. Have given it some considerable thought. |

## Test 3

### October 28, 1937.    11:30–12:00 P.M.

Back in the study of my own home at 380 Riverside Drive.

I have been refusing to think about Sir Hubert Wilkins during the day . . . holding thought impulses until the periods set aside for our attempted telepathic communications . . . so that my conscious mind will not "color" impressions by suggesting things which have occurred to me earlier in the day.

Suggestibility of the conscious mind and activity of the faculty of imagination—both of these factors are strong—and must be controlled for true receptivity.

The following is an exact transcription of impressions received:

| SHERMAN | WILKINS |
|---|---|
| *Unsettled conditions* | |
| *Weather still mild . . .* | Took train from Montreal and arrived |
| *Work progressing slowly* | Ottawa 1 A.M. |
| *You back Winnipeg today* | |
| *Russian fliers down 190 miles this side of Pole—plane wrecked—men dead—* | Have received several letters from people saying flyers about 200 miles from |

| SHERMAN | WILKINS |
|---|---|
| *think you'll find them this time . . . Don't know why I write this—except I seem to see or feel condition—Plane seems to have turned over on its back—propeller smashed—now half buried in snow and drift ice . . .* | Pole, two dead—Levanevsky and another. Also, many say 2 of the men are now walking away from plane for help, leaving the others there. Letters say plane right side up, wheels smashed and body of plane wrecked. Two people have visioned the men using the engine props loosened from shaft as windmill to drive motor and charge their batteries. |
| *You conferred 3 important people in Ottawa regarding flight . . .* | This okay. McHowe, Minister of Transport, Deputy Minister of Transport, Major Edwards, Signal Corps. |
| *Strange mental feeling tonight—I seem unable form words in dark—confused force—a mental pull of some sort—almost like static or interference—Like vibrations from Russian flyers—as though they knew we trying to communicate and trying to help from another plane of consciousness . . .* | |
| *79—133—where would this be?—These numbers come to me forcefully—as though spoken—I never felt this way before . . .* | About 300 miles SSE from where the Russians were last heard from. Possible they might be there. |
| *You did some business with man of reddish brown hair—blue eyes—see such a face vividly—genial yet penetrating . . .* | Major Edwards, rather thin faced. |
| *Delay a week—perhaps two* | Take one week to get radio from Marconi. |
| *C is with you—carries good luck charm . . .* | Cheeseman joined expedition this day. Carries a penguin (small, wood) as charm. |
| *Skis about fitted—still trouble or wonderment about weight on them.* | Skis were actually fitted to plane that day. |

## Comments on Third Telepathic Test

Some of the sensations received in this telepathic experimentation are impossible to describe, and they must be experienced by others in order to be fully comprehended.

Apparently, as soon as one learns how to blank the conscious mind, a new inner mental world is opened up, the boundaries of which have never been charted.

At times, for fleeting instances, I felt myself in Winnipeg—then again, out in the Arctic wastes near the position of the lost Russian flyers. In fact, it seemed, during most of this period, as though some force was trying to *pull* me out into the Arctic. This mental pull is a strange, and almost startling sensation.

During other moments of this period, I was conscious of being in my study, and then it seemed as though Sir Hubert Wilkins was standing beside me. Either that was true, or we had met one another at some unidentified spot in space between here and his location. Sometimes, too, I seemed to be moving about with him, contacting people and problems with him—as though I had tuned-in on what had been taking place during the day.

I had resolved to record my actual impressions at all times, no matter how illogical they might seem to my conscious mind or to others upon examination. To me, this is of the greatest importance in such experimentation.

I had not tried to familiarize my conscious mind with geographical locations. I wanted to keep as free from "coloring" as humanly possible.

I either saw or felt or inwardly heard the impressions which came . . . and I wrote them all down in darkness. Occasionally it was as though I was recalling something I already knew . . . at other times there was a quick positive flash as though I was receiving at that instant. More about my other reactions will be given later.

## Test 4

November 1, 1937.    11:30–12:00 P.M.

*My wife and I decided to take in a movie, and went down to the Capitol Theatre, at 50th and Broadway, where* Double Wedding *was playing, starring Myrna Loy and William Powell. We had intended to be home by 11:30 P.M., but the picture ran longer than we had contemplated.*

*At exactly 11 P.M., while engrossed in the picture, I had a sudden strong thought of Sir Hubert Wilkins—and, as we were leaving the theatre, at 11:29 P.M., I had a flash of certain numbers.*

Borrowing a little pocket notepad from my wife, I made notes of my impressions all the way home, as hereinafter recorded, arriving back at our apartment and going immediately into my study to complete notations, a minute or so after midnight.

The following is an exact transcription:

| SHERMAN | WILKINS |
|---|---|
| *(Following impressions received enroute home on subway train at appointed time) Strong thought Wilkins at 11 P.M. (while at Capitol Theatre) as though he keeping track time for appointment. 11:39—traveling sensation (unrelated to subway). You talk someone—then think of me during conversation—Traveling sensation again . . .* | Arriving Winnipeg 1:40 P.M. At Dinner Party with rowdy crowd until 11 P.M., your time. |
| *You communicated Russian government today regarding some flight matter—weather still mild—you advise cannot take off on search under two weeks unless radical change in weather . . .* | Talked to Embassy at 7:30 by phone to Washington—gave estimated time of departure for north as two weeks. |
| *Skis fitted* | Skis had been fitted during my absence. Wheels replaced on machine late tonight. |
| *Wheel comes to mind—cannot figure connection—wheel . . .* | Wheels placed on machine by working late at hangar. |
| *Your wife—something you want to tell her . . .* | Talked of my wife—she thought she might come here under engagement to sing—after meeting these people who interested in her, mentally decided to tell her not to come. |
| *You are in a place near a lot of people—I hear music—talk—"Oh, yes" I hear you say to someone, as though trying to beg off from engagement, "Well, not tonight . . ."* | Much talk—no music—Repeatedly refused to return to a house where wild party in progress. |
| | Five women—all "gaga"—four men in group. Wife of host very alert, energetic, |

| SHERMAN | WILKINS |
|---|---|
| | psychic (she thinks) *wearing green dress* (rather stout). |
| *You now talking animatedly with group—several women present—one very alert, persistent type—much interested in what you are saying—wearing green dress.* | Met and talked midnight, your time, with another woman, Kathleen Shackleton, artist, her brother—the explorer—Sir Ernest Shackleton. She also very positive, Irish, psychic. She drawing charcoal portrait of me. Also wearing *green dress*. Large woman, black hair, large face but not stout. |

# Test 5

## November 2, 1937.  11:30–12:00 P.M.

*I retired to my study a few minutes before the appointed time and tried to make my mind receptive to impressions. I found this to be more difficult than ordinary due to "excitement in the air."*

*I can only describe the mental sensation by comparing it to a radio at a time when many stations are trying to crowd in on the same wave length. It seemed as though there were countless vibrations or impressions in the background.*

*In my inner eye, I would see a myriad of sharp light particles against a field of black and moving crisscross shapes, threatening to take form—but it all remained a jumble . . . and throughout this sitting such a condition would return off and on.*

*Between times I seemed to establish actual telepathic contact as usual.*

*The exact transcription of my impressions follows:*

| SHERMAN | WILKINS |
|---|---|
| *Turmoil of election night—Feel excitement vibrations—Hard time getting you tonight—Confused directional sensation, as though you several places in short space of time . . .* | At a dinner given by Press Club of Winnipeg. Gave talk lasting until 11 P.M., your time. Then went to hotel. At 11:30, your time, was at home of Capt. Innes Taylor, ESC, navy man, who was with Byrd in Antarctic. |
| *You consult high naval officer—man experienced in north—drift ice—ocean depths—air currents . . .* | Discussing with Innes Taylor North and South ice conditions and work in ocean depths—the science of oceanography. |

| SHERMAN | WILKINS |
|---|---|
| | Unfortunately no chance of thought concentration on you. People talking and asking questions all hours until 2:30 A.M., your time. |
| *You have a pen or pencil in hand and are making notations or numerals . . .* | |
| *I am in a plane—flying low over sparsely wooded area—water—scattered houses—a three story building near waterfront—the plane lands on a field that has gradual rise of ground at one end—near water . . .* | We flew plane on wheels today for training second pilot and wireless tests. Country flat and sparsely wooded.

Town is beside river on flat ground. |
| *I see or feel 11 peole at field to meet plane—2 of them directly concerned your trip and seeing you . . .* | A great crowd of people (for Winnipeg) about 200—watching us in flight today. No new individuals personally concerned but many interested people. |
| *I feel your mind is disturbed about something—slowness of others to move or give cooperation actually required for quick action—Necessity of covering same ground twice to be sure done right . . .* | This much emphasized talk at Press Club dinner. |
| *You plan retire fairly early if possible for good night's rest and early rising as heavy day Wednesday.* | *Did* plan "early to bed" but had to go out to Innes Taylor's, and, although very sleepy, stayed up until 2:30 A.M. |

## Test 6

### November 4, 1937.   11:30–12:00 P.M.

*A half hour before the agreed-upon period tonight, while listening to the radio, I seemed to receive strong mental flashes of Sir Hubert Wilkins—as though he were actually thinking of me.*

*I tried not to open my consciousness to these flashes until the proper time, because I did not wish conscious recollections of whatever might have been received to color any impressions which might come later.*

*I am convinced that many telepathic impressions are being received by humans without their realizing it—and in the course of the day's activity. The conscious mind ordinarily rejects all such impressions as figments of the imagination, or one's own fear, or worry, or desire thoughts. In many cases, of course, these causes are responsible. A sifting out process is necessary—the developed ability inwardly to recognize the difference between impulses—whether self-created or originating from without.*

*Here is the exact transcription of my impressions which reveals confusion on numerical location. I am not permitting my conscious mind to check me, even though it might try to say, "that's wrong" . . . recording these mental pictures or impulses just as they appear in consciousness.*

| SHERMAN | WILKINS |
|---|---|
| *Have feeling I am to hear from you shortly—you are writing—Conditions more settled tonight—calmer—less disturbed sensation . . .* | Had rather worrying evening account of one radio man being refused permission to enter U.S. Had to telephone and wire several people, this country and Canada. |
| *"Work complete" I hear some voice say. "Ready to move on . . . need more snow."* | Made night flight with machine which was last part of training. All ready except waiting for radio man with equipment from Ottawa. |
| *De-icing equipment—need of same . . .* | Plane iced up badly on one flight during this night. Thought of adding to de-icing equipment. Now we have only de-icers on propellers. |
| *Death of some friend affects you . . .* *Feel you much closer to real action and more satisfied in mind . . .* | Some real action when found that pilot had, in my absence, taken three women up in plane. Because of great risk (unnecessary risk and bad judgment), fined him one month's salary. |
| *Hotel room—you seem to be in your room—fourth floor . . . Picture on wall—early settlers.* | My room in hotel 4th floor—413. Picture on wall, "Jesus" explaining the word of God to five old men, an open book on the knees of one man. |

# Test 7

November 8, 1937.    11:30–12:00 P.M.

*No confirmation, as yet, or check on impressions already received.*

Interference of the conscious mind, long trained as the conveyor of impressions concerning the outside world, gained through one or more of the five senses, is the most difficult to control.

Necessity of entering upon each session with quiet confidence that contact can be made—the stilling of all feelings and thoughts of doubt or wonderment or anxiety.

The following is an exact transcription of what came to me:

| SHERMAN | WILKINS |
|---|---|
| *Restless—weather discouraging—delay again strong impression . . .* | Wired this day to Canadian authorities and Russian meteorologist at Fairbanks for week advance weather forecast, since we were about ready; and was debating in mind whether I should go via Alaska or Mackenzie River. |
| *You turn some instrument in your hand—is it a range finder? At any rate, this is what comes to me . . .* | Located and bought this day a prismatic compass used for survey work. Had some difficulty finding it, so naturally gave it intensive thought. Three firms were instrumental in locating the compass at a second-hand shop. |
| *Barrow—this word came to me after I first saw mental picture of a wheelbarrow—Now I see a long pointer—and draw inference "Point Barrow". . . Are you going there for some reason—is there snow there? Have you decided shift operations? Where weather can be more immediately favorable? . . .* | Refer to first comment. Possibility of going via Alaska because of warm weather in Canada. This also received intensive thought during day—and Barrow was often in mind. |
| *"7"—this number came as though you about send new location . . .* | At an "evening" someone telling my "fortune" with cards. No. 7 was referred to repeatedly. |
| *Cough—is someone still bothered? . . .* | Had slight effect of head cold that day which I cured rapidly by application of "Kymol." |
| *You keeping time free this experiment still very difficult.* | This night thought of you and discussed experiment but "in company" had little time for direct "conscious" concentration. |

## Test 8

### November 9, 1937.    11:30–12:00 P.M.

*I confess tonight to having had a struggle to make my mind receptive. The very natural wonderment as to the degree of success or failure in these experiments, awaiting word from Sir Hubert Wilkins, made me overly sensitive.*

*Elements of doubt seem to upset the intricate or highly reactionary inner machinery of the mind. It required an effort of will, then a resolute "letting go," in order to clear the "screen."*

*The following impressions then came to me:*

| SHERMAN | WILKINS |
|---|---|
| *Canvas tent—do you have a portable unit of some kind? Newest equipment in case necessary camp on polar ice? I see some such arrangement packed,* with poles, and so forth, *in compact elongated bag to be stored in plane . . .* | Bought 8 long bamboo poles to be used as poles for tent which we shall use on the ice if forced down. These poles were difficult to locate. I examined several types of wood poles before finally getting word that bamboo poles were available. |
| *You received some money today . . . I see you get cash—you are in a bank.* | On November 6th, went into bank to ask if they would take check and refer it to New York, and then expected reply and money delivered to me on *Tuesday, 9th.* The banker, however, handed over the cash. The "thought," however, was intensely for delivery of money on the *9th.* |

## Test 9

### November 11, 1937.    11:30–12:00 P.M.

*Late this afternoon I received word from the New York Hospital that my dear friend, Charles E. Whitmore, was to be operated on at 7:30 P.M.*

*This necessitated my going at once to the City Club, and notifying certain of his friends and relatives, and then standing by to report on the outcome of the operation.*

I went out to the New York Hospital, then back to the City Club, phoning Mr. Whitmore's brother twice in Boston.

I just arrived back at my home, 380 Riverside Drive, in time to enter my study, and prepare myself for the test period at 11:30 P.M.

It was naturally more difficult than usual to clear my mind of the intense personal events of the past few hours—but, if accurate results may have still been obtained, then this experience under stress will have been of definite value.

Here is an exact transcription of my impressions:

SHERMAN

*You at Winnipeg—Busy day—You roped in on Armistice observance—Tribute to Canadian war dead—Flowers dropped from plane . . .*

*"Aircraft in next world war to annihilate civilian population." You in plea for constructive use of airplane and modern inventions—great forces to be liberated for use of humanity—Tribute fellowship aviators of all nations—fact you going on expedition to search for Russian flyers—to attempt to do for them as you would wish to be done by—were you, of a different nationality, down in the Arctic wastes . . . You bemoan loss of fellowship between nations when so great a work to be done by all . . .*
*These thoughts running through your mind if not actually uttered in Armistice address.*

*You in company men in military attire—some women, evening dress—social occasion—important people present—much conversation—You appear to be in evening dress yourself . . .*

WILKINS

We flew past Regina almost to Saskatchewan, but, account of bad weather, returned to Regina and landed at 12:30. As we flew over Regina a service was being held at the Cenotaph. We were directly over it, but did not drop flowers.

These thoughts spoken at breakfast gathering when entertaining mayor of city, counsellors, and others just before leaving Winnipeg at 10 A.M.

Armistice Ball at Regina. Many officers of army and police in uniform. Had "tea" with Lieutenant Governor of Province, and supper with him during the evening. *My appearance at this affair was made possible by the loan to me of evening dress.* I hesitated about accepting it, and gave it intensive thought. I *did* wear it, however.

| SHERMAN | WILKINS |
|---|---|
| *Something mechanical doesn't suit—you glad it acting up now—rather than later—De-icing—serious consideration of more extensive equipment—feeling strong here you should watch this for own protection . . .* | Decided to fit, and installed on the plane, not a de-icer, but a nose cowling to keep the engines warm. |
| *Someone seems to put or pin something on your coat lapel—either pins a medal on or token of some kind—I hear you say, "on behalf of . . ." and rest is lost . . .* | At the "breakfast" occasion at Winnipeg, the mayor presented me with a "city badge" and the freedom of Winnipeg. *The badge was actually pinned on my coat lapel by Mrs. Innes Taylor, wife of one of the men who was with Byrd in the Antarctic.* |
| *You pleased with charcoal likeness—seem to see you facing toward my left as I look at picture—about 3/4 front view . . .* | On this day was published in several newspapers a copy of a charcoal "likeness" drawn by Miss Shackleton. I saw the reproduction of course, and gave it some thought. |
| *"Phillips"—were you talking some man by this name? He seems to offer you cigar . . .* | This morning a Mr. Coyne, don't know his first name, presented me with a box of twenty-five cigars. |
| *Some slight friction at certain points—See you "telling some man off" quite emphatically—give him to understand, "if this happens again, you're through with this expedition"—Impression—drink and women and insubordination.* | The "telling off" occurred, and there is some substance to this: "drink and women and insubordination," but not on this particular day. The possibility of carrying out the dismissal is often strongly in mind. |

## Test 10 (Wilkins at Edmonton)

November 15, 1937.  11:30–12:00 P.M.

*A headache which persisted through this evening caused me concern as to its possible interference with my receiving of impressions at the appointed time.*
 *In an attempt to gain relief, I lay down and dozed, but got up at 11:15 P.M., and went into my study, still plagued by the headache.*
 *In addition, having dropped off to sleep with my thoughts on Wilkins, I had*

returned to consciousness with the vague sense of having been far away. This is a difficult sensation to describe—except to say that one does not feel "all there"—and it took an effort of will to "pull myself together."

The following is an exact transcription of what seemed to come through.

| SHERMAN | WILKINS |
|---|---|
| *Get third floor impression—as though you so located—or in room at present time. Skis put back . . .* | My room on third floor. Skis fitted this day. |

## Test 11

### November 16, 1937.   11:30–12:00 P.M.

*I felt particularly keen for the telepathic experiment tonight, and looked forward to the time with real anticipation.*

*I would occasionally get flashes when my mind seemed to traverse the distance between New York and Edmonton—making contact with Wilkins' activities there. I did not try to "pull these flashes over into my conscious mind" until the hour arrived for our telepathic test. Had I done so, it would have left a half-completed recollection of an impression which might have become colored by my imaginative faculty.*

*One's imagination is always one of the obstructions to be guarded against.*

*The following is an exact transcription of the impressions received:*

| SHERMAN | WILKINS |
|---|---|
| *Entertained by Edmonton Men's Club—like a Rotary organization—Are greeted by former friends and acquaintances—I hear someone referred to as "Lord Mayor"—he is welcoming you and crew, and wishing you "Godspeed" . . .* | Probably, not quite sure of date. |
| *You in hardware store for some reason today—seem to see you buying strands of wire—and other odds and ends—Store you are in seems to have a veranda or porch along front—double row of counters—a girl clerk and three men . . .* | Hudsons Bay Store—Bought variety of equipment. A girl clerk did a good job of phoning all wholesale houses even as far as Toronto in search of a special compass for me. |
| *See you called to long distance phone—* | |

| SHERMAN | WILKINS |
|---|---|
| *Russian Embassy—also Ottawa—some sort of clearance provision—rights to use some base of operation—secure needed cooperation to get supplies through . . .* | Very likely on this date. Soviet Embassy phoned about this date. |
| *Artificial flowers—someone has made some exquisite ones which have come under observation—can these have been made by a blind ex-service man? Something made by hands of person handicapped in life seems to have been presented to you . . .* | A blind ex-service man came to me. He represented a life insurance company. He proposed that his company insure me for a large sum, the policy—if I did not return from this trip—to go to the inauguration of a Wilkins Polar Science fund. |

# Test 12

## November 18, 1937.  11:30–12:00 P.M.

*During the day I several times thought strongly of Sir Hubert Wilkins, as though his mind had been directed toward me. I have not tried to pull through impressions on such occasions, preferring to leave them in my subconscious, and let them come through at the appointed times.*

*The exact transcription of my impressions received tonight is as follows:*

| SHERMAN | WILKINS |
|---|---|
| *I see you in plane seated beside first pilot . . .* | A full-load flight test today intending start tomorrow. |
| *Things have been more hectic, if possible, than at Winnipeg . . .* | Sat up all night writing reports and letters. |
| *One of expedition seems sick—something he's eaten—"W"—Is it Wilson? . . .* | Wilson was sick today, also yesterday, but is recovering. |
| *Plane being tested full load—this fairly hazardous—one rather close call—take-off . . .* | Plane was tested this day with full load. |
| *9:30 tomorrow morning you plan some definite action—perhaps resume flight—something important set for that time . . .* | Left word by letter to all of expedition staff to be at airport at 9 A.M.—Take off at 9:30. |

| SHERMAN | WILKINS |
|---|---|
| *My letter addressed to you at Winnipeg has reached you in Edmonton—you plan reply before you are beyond point of contact, busy as you are.* | Letter did reach me that day—replied that night. |

## Test 13

### November 22, 1937.   11:30–12:00 P.M.

    *I made arrangements by phone with Reginald Iversen, radio operator for the New York Times, this last Sunday night, to forward condensed reports my impressions to Wilkins in far North, when expedition has wireless set up.*
    *Iversen asked me to report to him if I had any impression Wilkins' radio was ready to communicate with him before so notified by wire by Wilkins. Iversen said, should I receive such an impression, he would go on air with a "call" for Wilkins prior to hearing from him—suggesting this would provide another telepathic test.*
    *I received impression regarding wireless at this "sitting," as hereinafter noted. The following is an exact transcription of what "came through."*

| SHERMAN | WILKINS |
|---|---|
| *Following Mackenzie River in flight— weather—fog and snow—further delay —down at town with old stone fort— expect fly on tomorrow morning— Aklavik goal . . .* | Followed Mackenzie River from *Fort* Resolution to Aklavik this day (22nd), flew through snow and fog. |
| *Work on plane—radio not perfected satisfactorily yet for service long flight . . .* | Much concentration on radio in plane. |
| *You and crew feted by townspeople— Inn-like place . . .* | Dinner party at the Village Inn. |
| *Several uneasy moments today in air and on ground . . .* | Bad flying—pilot wanted turn back but, I said, "try ahead" and we got through. |
| *Another plane following—it takes on additional supplies . . .* | The second plane, with supplies, was in my thoughts. It was twice forced back to Resolution by bad weather, but we flew through the weather. |

|  SHERMAN | WILKINS |
|---|---|
| *French-Canadian—a man of French descent—talks to you quite excitedly or interestedly about something—tries to convince you a certain thing is all right—you shake your head good-naturedly but firmly—he is looking at your plane—Later you have meal with this man and some friends of his—someone gives you some delicacies to eat—remarks about your soon not having chance for this kind of food . . .* | All okay. Stayed in house of French-Canadian doctor at Resolution, 20th, 21st, and left on 22nd. |
| *Color white flashes—followed by blue—am trying to get significance blue which mind ruled out few minutes ago—something having blue color associated you this day . . .* | I am wearing a dark blue parka. |
| *Feel sense impending action after delay or stop-over on flight north from Edmonton.* | We were delayed at Resolution one day, and left on the 22nd. |

## Test 14

### November 23, 1937.   11:30–12:00 P.M.

Ten minutes before time for the experiment this evening, Edward Reese, an actor friend, phoned. He was full of a number of activities in which I would normally have been interested and ready to talk at length.

Now, my mind already seemed to be making "contact"—there were thoughts and impressions that seemed to be pressing against the conscious mind, waiting to "break through," and I paid more heed to them than I did to what was being said on the phone.

Finally telling Ned Reese that I had to finish up a writing job, I got off the phone at 11:29, and seated myself at once, switched out the lights, and blanked my conscious mind.

The following is an exact transcription of impressions received.

| SHERMAN | WILKINS |
|---|---|
| *Eager feeling as though you thinking of me few minutes before time our attempted communication . . .* | Was thinking of you after arrival at Aklavik. |
| *You Aklavik—been there over 24 hours—Vibrant quality in air—different attitude of mind—a vital concentrated radiation from you . . .* | We arrived Aklavik yesterday. (Curious that you should sense Aklavik this day—23rd—when on the 22nd you thought we would get there on Wednesday, the 24th!) |
| *Guns—did you try them out? Target practice . . .* | I unloaded the plane and in evening cleaned both guns, and selected the one I will take with me, the other one I will leave behind. |
| *You seem to be in three-story building—men's rooms down hall . . .* | We are making our headquarters in a three-story (including the attic) Road House. The men's rooms are adjacent, down the hall. |
| *Feel you came through okay from Edmonton—see you down twice—you make Aklavik in two hops—plane performs well . . .* | We started from Edmonton on the 19th, but had to return. Left again on the 20th and reached Resolution on the same day. |
| *Wireless tests New York being readied—appears to have been blustery day . . .* | *Was* blustery day. Bad weather. Our supply plane started out twice, and was forced back each time. |
| *You've been to dinner with some officials—men and women, several children in teens present . . .* | Had dinner with the government doctor. He has a little girl, and keeps an "adopted" Eskimo girl as company for his daughter. |
| *Busy day—checking over everything—getting base established for all operations . . .* | Okay. |
| *Some member of crew in none too good humor over something—seems to be working with a grouch—you disregard attitude—letting wear off . . .* | Okay. |

| SHERMAN | WILKINS |
|---|---|
| *Snowshoes—I seem to see you examining a pair.* | Okay. |

## Test 15

## November 25, 1937.   11:30–12:00 P.M.

*On Wednesday afternoon, through the kindness of Dr. E. Stagg Whitin, I had an interview arranged with Dr. Gardner Murphy of Columbia University, who said he would be glad to check these experiments.*

*I am to mail a copy of my impressions each following morning to Dr. Murphy at Columbia, who will read them, and file them away.*

*Dr. Murphy has suggested additional tests, such as the mental broadcasting by Wilkins of numbers and symbols, along the lines of the Duke Experiments. It is Dr. Murphy's feeling that my ability to pick up definite numbers or symbols or letters of the alphabet at this long distance will serve to further substantiate the other impressions I am receiving of Sir Hubert's activities.*

*I find it is necessary to adjust the conscious mind to each new condition. Tonight, with the realization that Dr. Murphy is to check each succeeding night of impressions, it was difficult to rule out a mental element of "trying too hard to pull in impressions." No doubt, as I get better acquainted with Dr. Murphy and his associates, this tendency toward "self-conscious" effort will be overcome.*

*The exact transcription of my impressions received is as follows:*

| SHERMAN | WILKINS |
|---|---|
| *You charting planned flights today—going over maps, discussing with crew and some man, weather and topography expert . . .* | Okay. |
| *I sense air activity of some sort, not related actual search flight—in preparation therefore, but not in search plane . . .* | This contemplated, and at that time desired. |
| *Have dinner with group—all talking agitatedly—"But how about this, Sir Hubert," and "What do you intend to do if . . ." I hear questions asked . . .* | Besides other Road House boarders, three Catholic priests from Catholic Mission, had dinner with us. They all with long Arctic experience, questioned me closely on all matters, especially the submarine. |

| SHERMAN | WILKINS |
|---|---|
| *You leave this group talking with K and go to your room . . .* | I was very tired, and left the "party" early. |
| *Somebody offers you home-made candy which you can't very gracefully refuse . . .* | Home-made candy at dinner, and handed around later. |

## Letter to Sir Hubert Wilkins at Aklavik

November 26, 1937.

Dear Sir Hubert:

It was fine of you, despite the pressure of your duties, to get off the second batch of impressions to me, with your marginal notes. They were even more encouraging than the first, in my opinion, containing several quite remarkable instances of my apparent picking up of "thought forms."

Acting upon your permission to seek the cooperation of Dr. Gardner Murphy of Columbia University, I am glad to report having seen him, and his having volunteered to work closely with us.

Dr. Murphy has suggested that I send you a pack of Dr. Rhine's ESP cards, containing the symbols (heavily outlined in black) of a square, a circle, a cross, a wavy line (really three wavy lines in a perpendicular position), and a star. It is his wish that we experiment in the sending and receiving of a certain number of these symbols . . . or numbers selected by you . . . or letters of the alphabet. Should I be successful in receiving a goodly percentage of these impressions, in addition to accurate mental pictures of what you are experiencing, Dr. Murphy feels that we will have built up a strong case for extrasensory perception.

What I will have to guard against, during this next period, will be a "self-consciousness," "a trying too hard to pull impressions in," realizing the test conditions set up, which I welcome, but which, nevertheless, must learn to "take in stride," naturally, and as relaxed as ever, mentally and physically.

Iversen at the *Times* said he'd be glad to radio condensed report of my impressions to you the following day . . . and receiving back your check on them. In the event we add to our experiment by your endeavoring to send me impressions of symbols or letters of the alphabet, picked at random . . . and later in actual spelling of words, then a system or rotation will have to be devised, so I can set my mind accordingly. It would be easy to radio impressions of such a test back and forth, saving my mental picture impressions for confirmation by mail.

I will appreciate your suggestion as to what you wish to attempt to send from your end . . . and I will do my best to cooperate. Iversen will communicate with me if you advise him. My best wishes always—

## Test 16

### Novevmber 29, 1937.   11:30–12:00 P.M.

*No word now for some little time from Sir Hubert Wilkins—since receipt of his letter, and notated copy of my previous impressions on November 21st.*

*No news of his activities in the* New York Times—*so no way to gain any knowledge as to correctness of current impressions.*

*I personally prefer having no conscious awareness of what is going on—for I find that an occasional news item which comes to my attention, though it may confirm some impressions received, also tends to inhibit me in bringing additional impressions through.*

*Fortunately there has been little in the papers concerning Sir Hubert Wilkins thus far. Publication of any stories always occurs after my impressions of a day's events have been received. In the few instances that I have seen news accounts, they have seemed at variance with my own impressions as might concern personal details, until I have heard from Sir Hubert Wilkins himself, and learned, to my own surprise, how accurate my impressions have been.*

*This observation indicates how little one's conscious mind can be depended upon as a guide in telepathic communication. Dependence upon it, in any way, always produces "coloring" or imaginative excursions—and stops the flow of genuine thought waves.*

*The following is an exact transcription of impressions received tonight:*

| SHERMAN | WILKINS |
|---|---|
| *You come back from flight and remove instruments from plane for adjustment . . .* | This on 27th. |
| *A padre or priest calls on you—he is friendly and has shown you several kindnesses before . . .* | This on 25th. |
| *I see you some place where there is oil lamp—it is carried about and casts a shadow—as you move behind or along with person who holds it . . .* | I changed oil lamp I had been using for a gas burning lamp, this day. The oil lamp gave a yellowish light, and I found it difficult to read by. |
| *Spirits of men good—though one man seems partially laid up or unable to do much at present—can't sense whether ill or unable be active due his work being temporarily done—he idle—sits around or looks on . . .* | Okay—*the usual one.* |

## Test 17

### November 30, 1937.   11:30–12:00 P.M.

*In tonight's sitting, what appears to have been a bit of "coloring" occurred.*

*I received a strong impression of "ping-pong balls," for some reason, and found myself writing: "sudden flash of ping pong—is there table in town where people play? Can't account this unusual impression . . ."*

*Following my practice of recording every impression, exactly as it comes, without permitting my conscious mind to interfere, I set the above down.*

*When the session was over and my conscious mind again in control, I was reminded that a "ping-pong ball" episode took place on the Richman-Merrill flight—and, as I recall, in another plane. The Wilkins-Soviet Expedition purchased the plane Merrill used in making the Coronation flight.*

*Now here, on the surface, appears to be one of the first evidences of an intrusion of past knowledge into my impressions. Unless Sir Hubert Wilkins has encountered the game of ping pong being played at Aklavik, or this incident actually happened, then this is a sample of what occasionally happens when events that might possess an associative value, crowd in from the memory channels of the mind.*

*Certainly, in most of my impressions received, there has been nothing associative to suggest a carry-over from one impression to another . . . the type of carry-overs so often produced when one's imaginative faculties are brought in play. I have learned, for the most part, how to "kill off" my imagination in making my mind receptive . . . and this absolutely must be done to receive outside impressions.*

*The exact transcription of my impressions is as follows:*

| SHERMAN | WILKINS |
|---|---|
| *Strong impression ping-pong balls—sudden flash of ping pong——is there table in town where people play? . . . Can't account this unusual impression . . .* | Two of the men, Cheeseman and Dyne were playing ping pong in the school gymnasium, but I was not there. |
| *Night work impression—something hauled on sleds to hangar or building nearby—workmen coming to and from place—busy on some job—carrying supplies or equipment . . . you on hand supervising . . .* | Working until late at night on radio shack placed out on river. |
| *Latitude 68*<br>*Longitude 133 . . .* | Latitude 68<br>Longitude 135. |

| SHERMAN | WILKINS |
|---|---|
| *Star reckoning—I see you looking at sky—thinking about several flight courses...* | Was taking observation of Pole star to fix the true north for radio beacon. |
| *Personnel not as happy a combination as it might be—and you wonder, at times, what outcome may be in one or two instances—are not too highly pleased with some of attitudes of one or two men—human element—you may yet decide to get new man and have him flown to Aklavik, to join party—are turning this over in mind as possible recommendation to Russian authorities...* | Okay—strong thought. |
| *Still feel some special activity Wednesday—tomorrow—work on wireless and tests to be continued and intensified—this is one of most important concerns now...* | We were hurrying in order to carry out first test with Iversen. This was successfully carried out on *Wednesday*. |

## Test 18

### December 2, 1937.    11:30–12:00 P.M.

Returning home from downtown tonight where I had attended a meeting of the American Society for Psychic Research, which was addressed by Dr. Henry S. W. Hardwicke, I found a phone call awaiting me from Reginald Iversen, New York Times *radio operator.*

It was almost eleven-thirty, time for my telepathic appointment with Sir Hubert Wilkins when I reached Mr. Iversen at his home. He asked me, testily, what impressions I had received lately, and I referred to my notes of the past several sessions. I told him of my impression that wireless communication was to be attempted between New York and Aklavik on Wednesday—and Mr. Iversen asked me if my recorded impression was then in the hands of Professor Murphy at Columbia University. I assured him that it was. Then he said: "Well, your impression is pretty good, because we started our wireless tests last night (Wednesday) around seven-thirty P.M."

These impressions as to the time came in my sittings of the past Monday and Tuesday nights. On Monday evening I recorded: "Working on wireless—testing...

Wednesday important action day—weather permitting." On Tuesday, my impression in this connection was: "Still feel some special activity Wednesday—tomorrow—work on wireless and tests to be continued and intensified . . . this one of most important concerns now . . ."

Mr. Iversen stated that the New York Times *was to carry quite a story of Wilkins' activities and plans in the Friday morning edition, but said he would tell me none of the details. I told him by no means to do so, as it was difficult enough to keep my mind from "coloring" without its being given certain known facts to elaborate upon.*

Mr. Iversen can testify that I never have asked him information concerning Wilkins in advance of any impressions received. He is keeping a careful record, of his own connection with this expedition, as an aid in checking the correctness of my impressions.

Herewith is an exact transcription of tonight's session:

| SHERMAN | WILKINS |
|---|---|
| *First search flight to take place Sunday or Monday—weather permitting . . .* | Flight to Barrow planned for Sunday or Monday depending upon weather. This was actually to be first leg of search flight. Sent request to forecasters for weather in relation to Barrow flight, so thought was "active." |
| *Crew carefully trained to meet all emergencies . . .* | Cheeseman was ordered to familiarize himself with controls, since he was to fly the plane to Barrow. |
| *You operating between a number of base locations—Point Barrow—Barter Island—Aklavik and a fourth location—seemingly still further north . . .* | Fourth station—Baillie Island—250 miles north-northeast of Aklavik. |
| *First flight intended jaunt 600 miles—out and back—covering area mathematically . . .* | Flight from Barrow planned for Sunday or Monday—distance—600 miles. |
| *Impression still persists of possible replacement of one of men—another pilot flying up to Aklavik to take someone's place—strange, last minute change some man in personnel seems imminent—either one man is forced to drop out because ill, or he doesn't fit as well as had been expected—feeling so strong if this hasn't happened, believe it will soon occur.* | Same impression—same man. |

## Letter to Sir Hubert Wilkins at Aklavik

December 3, 1937.

Dear Sir Hubert:

Again quite an unavoidable stretch when I have had no opportunity of learning degree of accuracy of my impressions.

As you will see by these records, I felt at first that you should be ready to start your wireless tests with Iversen in New York on Friday or Saturday of last week . . . but by Monday and Tuesday of this week I knew that this would not occur until Wednesday, and so noted. Now, whether this had been your earlier intention, changed because of mechanical delays, I cannot know until the check-back from you.

I have received a fairly true picture of your present location, as described in my past two weeks' impressions, including the snow banked against your shelters, and so forth, much as narrated by you in this morning's *Times* story.

Dr. Murphy has suggested, when you get so you can, that you have another man or men in your party shuffle the ESP cards—and give them to you, one at a time, for you to send impressions of the symbols to me, keeping a record of their order. If I am successful in picking up a goodly percentage of these broadcasts, in addition to the sensing of conditions, surroundings, and so forth, he feels these tests will have proven of great scientific value.

I should know, however, that you are actually sitting, so I can focus my mind with every degree of confidence and expectation. I would have to eliminate all such factors of conscious doubt or apprehension in order to clear my inner consciousness for an attempted pick-up of anything so absolutely accurate and specific as the figure of a star, or square, or cross, or wavy line, or circle.

A certain number of tries should be decided upon—and either impressions of a cross and circle, all of them shuffled together and selected . . . or the whole deck, so shuffled, broadcast, with every symbol in action. I leave that to you.

Best wishes! My face burned the other night as I was sitting . . . wish I'd recorded this impression. I wrote instead "weather biting."

Sincerely—

## Test 19

December 6, 1937.     11:30–12:00 P.M.

*Somehow I feel that this is least satisfactory of my sittings to date. A number of factors may have contributed—first, a rushing day with my being late at appointments* and even a few minutes late getting home for this test period. More difficulty than ordinarily, clearing mind of day's events which had pressed in on me.

Then, too, I was aware of article I had read of Wilkins' activities, written by him, and as impressions started to come I could not restrain a conscious wondering if they had not been inspired by my possible recollection of certain incidents in the Times *news report.*

For an ideal test condition over a period of time, I should be as isolated as Wilkins, and then I have conviction that results would be extremely unusual.

As it is, to obtain a fair percentage of right impressions, despite stress and strain of day's events—as I appear to have been doing—seems indicative that this telepathic quality of mind is not too hard to put in operation—and may under better control conditions—be developed to a high degree of effectiveness.

Apparently the less I know of what has actually happened, and the more I have to mentally "grope" or "feel" for knowledge concerning Wilkins, the more authentic are the impressions received.

I think this is because my conscious mind, ignorant of actual conditions, can then not combat my inner impressions by comparative facts or imagination based on previous information.

These reflections are important as they bear upon the mysterious operation or inoperation of this psychic or telepathic sense.

An exact transcription of tonight's impressions is as follows:

| SHERMAN | WILKINS |
|---|---|
| *Feel out on ice with you—Test flight with wireless tested while in air . . .* | Flying over ice to Barrow this day. Testing wireless while in flight to Barrow. |
| *Feel Iversen at New York Times working with you tonight—wireless reception—may try reach him there after session—seem to tune-in his mind, wherever he is—thinking your tests, our experiments—something unusual being done today . . .* | Barrow flight. |
| *Thought of Lincoln Ellsworth connection with you—did you communicate with him recently—possibly through Iversen? Something about future plans . . . Several messages you wish delivered. Seem to see Iversen phoning Lady Wilkins or writing her note . . .* | Ellsworth repeatedly requesting me to go South with him. Several communications last few days. I had not answered them, and thought frequently about this neglect.<br>(These impressions confirmed also by letter Iversen to Sherman. Iversen also stated he had both phoned and written Lady Wilkins re: Message from Sir |

| SHERMAN | WILKINS |
|---|---|
| | *Hubert and had been thinking strongly of Sherman during flight, wondering if Sherman was sensing it during regular sitting period.)* |
| *Camp on ice plenty cold . . . I seem to walk out about a block or more from shore line of river from place where plane located, to little settlement on ice.* | Plane located about 200 yards from house. Plane on shore line. We cross a lagoon—frozen, of course, to get to the house. Lagoon end would appear to be a *river*. Our house one in a little village or part of the village. The houses at Barrow are on both banks of a narrow end of a lagoon. Most of the houses are "shacks," one-roomed affairs resembling a garage. |

## Letter to Sir Hubert Wilkins at Aklavik

December 7, 1937.

Dear Sir Hubert:

Rather than waiting until after my Thursday night "sitting" as heretofore, and then forwarding you my week's total of impressions, I am dispatching the Monday and Tuesday night sessions to you following tonight's attempted telepathic communication.

I am doing this to be sure to get this letter and latest impressions through to you at Edmonton, and thence on to Aklavik by around the 15th of the month. You should have quite a batch of material from me by that time . . . and I should appreciate your advising Iversen when I may expect to receive it back with your customary marginal notes.

I hope your memory, plus the diaries you have been keeping, will enable you to check these impressions as accurately as before, when you were not quite so remote from mail service.

I have been thinking a great deal of Dr. Murphy's suggested tests, to be added to what we have already planned. I want to be certain we do not undertake too much at any one time of a "mechanical nature," however exact the designed result in the event of successful reception.

What I constantly have to guard against is any "straining" to get an impression, any "over-anxiety" . . . and trying to force a result or to imagine what a condition or number or symbol might be.

When you are ready to give real concentration to these tests, I suggest we give one night in three to an attempt to send and receive a certain order of the ESP cards out of a certain number to be selected by you . . . shuffled by one of your men . . . and handed to you, one by one, for visualizing and mentally broadcasting. I can try to receive impressions of the whole 25 in one deck, if you like, as selected by one or more of your men, and handed to you . . . you keeping a record of their rotation.

It has been suggested here, to avoid any possible questions of collusion . . . that you then wireless the rotation of these symbols as selected and visualized by you . . . viz: circle, star, wave, circle, and so forth . . . as the case may be . . . and I will then turn my list of impressions over to Dr. Murphy who can consult with Iversen, so that I have nothing to do with checking impressions.

Perhaps such elaborate precautions aren't necessary, and yet I want to be sure that a skeptical public is reasonably satisfied these experiments have been conducted in a strictly scientific manner, leaving no loophole for any challenging of accurate impressions received.

The few who have examined our records thus far state that the results have been amazing. I only know, when conditions seem right, I can receive vivid pictures and definite feelings of what has been happening to you. I have discovered, and I have since been told that Mesmer used this same method . . . that turning off the lights, and then turning on a flashlight and looking at it for an instant, serves to intensify my visualizing ability or my extrasensory faculties. After looking at the lighted-flashlight bulb, then snapping it off, I close my eyes or look straight ahead in the darkness . . . and, as the image of the light fades in consciousness—*a scene of what has happened or is happening to you, comes in!* Try this some time, concentrating your mind on some distant point as you do it—and note what occurs.

I am greatly interested in the human side of your activities—and I wish to continue to record these impressions in addition to doing just enough of these other experiments to afford an additional check.

Of course, the receiving of number impressions is equally, if not more important, since a developed accuracy in this line may prove most serviceable in the event that you were cut off from civilization through temporary failure of radio. When you get so you can—why don't you propose the nature of this type of experiment and I will cooperate.

Let me say here—it is my belief that it is easier for a "sensitive" to pick up impressions of what is happening to you than it is to receive correctly numerical or symbolic impressions. This is because you have an emotional reaction, often intense, to everything you experience . . . and this is not so in the mental broadcasting of numbers or symbols. Seemingly, an emotional impulse sends a thought form on its way with greater vitality, and it breaks through more readily. I think there is another reason—so many images of what has happened (of various phases of your experience) are consciously or unconsciously transmitted ... that if a "sensitive" misses some of them, he picks up others. In the case of a number, the chances of receiving it are twofold ... either through an auditory sensory perception or a visual faculty ... one either hears the number spoken, seemingly, or sees it loom up in the inner mind. Any variation therefrom means a wrong impression.

When sending to me—try different methods—speaking the symbol out loud as well as visualizing. You don't need to try to put what has happened to you in words ... unless you want to mention name of specific member of party or concentrate on first letter of last name ... then convey impression of what may have happened.

This whole field of exploration is so new, we will have to develop methods of procedure as we go ... but I feel now that results to date indicate the possibility of a significant pioneering achievement well worth the investment of our time.

Perhaps you would like to attempt to receive from 12:00 to 12:30 some nights that we are working—instead of just sending. If so, notify Iversen and tell him the nature of the information you would like me to "broadcast," and keep a record of your impressions to send to me.

I have not met the other men associated with you, but will you convey my deepest good wishes for a happy, successful Christmas and New Year in the far North!

My most cordial personal regards to you whom I sometimes feel I can reach out and almost touch. Sincerely,

## Test 20

December 7, 1937.    11:30–12:00 P.M.

*Tonight, prior to my appointed time, I spent more than an hour writing a long letter to Sir Hubert Wilkins.*

*Whether this concentration upon Sir Hubert was responsible for the flow of my impressions, or whether I was particularly in the mood for tonight's session is something which will take further experimentation to determine.*

*Most certainly, more than at any other sitting, I felt such an urge to write and keep writing, that it was difficult for me to keep abreast of impressions or description of mental pictures or feelings which swarmed in upon me.*

*I shall await, with great interest, particularly, Sir Hubert's check of the following impressions which I have copied exactly, as customary, from the notebook in which they were originally recorded:*

## SHERMAN

*Don't know why, but I seem to see crackling fire shining out in darkness of Aklavik—get a definite fire impression as though house burning—you can see it from your location on ice—I first thought fire on ice near your tent, but impression persists it is white house burning, and quite a crowd gathered around it—people runing or hurrying toward flames—bitter cold—stiff breeze blowing . . .*

*Your plane looks like a silvery ghost in moonlight—I seem to be almost under nose of it—standing in snow, looking up—it towers over me—I've never seen plane of course, but it seems to have high bow, with two huge propellers either side of cabin or cockpits—motor concealed in great silver metal tube-like cylinders or encasements—don't know technical names for purposes of description—a rounded metal door seems to lift up to admit entrance to cockpits from top of plane—big instrument board front cockpit—seats for two—pilot and co-pilot or navigator—rear cockpit and space beyond for another passenger and storage—separate rounded metal door with glassed-windows covering each cockpit—plane rests on giant skis—dark in color . . .*

*Windows in plane seem of slit nature—wide panels that go around front and sides—as well as observation opportunity through top of plane—even though doors clamped shut—Everything fastened down—compactly packed, plane not yet equipped or stored with things as it will be on actual search hops . . .*

*17th comes to mind as real take-off for search flight—day later than you had originally*

## WILKINS

While I was in Radio office at *Point Barrow,* the fire alarm rang. A long ring on the telephone. (There are only four telephones at Barrow.) It was an Eskimo's shack on fire. The chimney blazed up, and the roof took fire, but it was soon put out. Some damage resulted, mostly from the efforts of the zealous firemen. Was pretty cold that night with a light wind.

Description of plane *practically exact.*

Okay, as previously stated.

| SHERMAN | WILKINS |
|---|---|
| *contemplated—this different sort impression involving time dimension—seem foresee unfavorable weather conditions arising to prevent action on 16th...* | Weather was and remained unfavorable over whole moonlight period, 15th to 17th, inclusive. |
| *Lights in town on high, thin poles—cast weird reflection over none-too attractive landscape—with frozen spots of what seems to be lake or water areas back beyond town...* | Several electric lights at Barrow on high thin poles. One on high mast is kept burning constantly as guide to trappers and travellers Barrow-bound. It is claimed that this light may be seen from a distance of thirty miles, if the air is clear. |
| *There is a road that seems to run a little distance back from river bank—making a turn around curvature of land as river seems to widen out beyond Aklavik into quite a bay—which stretches on and on as far as eye can see—this is main road in Aklavik—most of houses and stores of any consequence along it...* | This is more or less a description of the lagoon, with the river-like entrance at Barrow. |
| *A prominent citizen in Aklavik has died—and I seem catch glimpse of funeral service—strange sensation—this connection that an Aklavik doctor is also an undertaker—or somehow associated...* | There was a funeral service. A baby died (Eskimo). The natives act as their own undertakers. |
| *Is there some man you deal with in Aklavik by name of Webb or Weber? Name comes to me, and man seems medium height, heavy set, heavily clad—hooded garb.* | An owner of a store at Aklavik is PEFFER—about as described. |

# Test 21

## December 9, 1937.    11:30–12:00 P.M.

I could not escape what seemed, for the first time, a rather routine feeling in connection with Wilkins and his expedition—as I sat down to clear my mind for the session tonight.

It seemed that Wilkins had things pretty well under control, and almost ready for the actual search—so that from now on it is practically a case of "standing by" for right conditions in which to take-off.

Whether this feeling has inhibited my impressions tonight remains to be seen. I have experimented of late with sending my mind out, inquiringly, to see how accurately I can pull in mental pictures of Wilkins' physical surroundings . . . the town of Aklavik, and now, Point Barrow. How much of these impressions I receive from the subconscious of Wilkins, and how much of them are picked up clairvoyantly is difficult to determine—if they prove to have been right.

An exact transcription of the impressions received tonight is as follows:

| SHERMAN | WILKINS |
|---|---|
| *Point Barrow—some kind of banquet—yourself and crew guests—seem to see it held in church—women of church serve meal—talks—minister—town official—some manager Hudsons Bay Trading Post—you tell something your plans for search Russian flyers . . .* | Banquet at Missionary's House. His wife served meal to fourteen of us. Most all the "eligible" whites in the village present. (No "H.B." Post at Barrow) |
| *You sit next to man with dark robe and decorative wide lapels—either clerical or lord mayor uniform—this is quite an occasion for Point Barrow.* | The Missionary. |
| *May be coloring—but I see you connection school—standing front of blackboard—chalk in hand—you give short talk, illustrating remarks . . .* | I gave a talk to the school children today. |
| *. . . testing conditions—seem to hear you say, "Hello, Iversen—Wilkins speaking—come in . . . call numbers SN-2147X." Don't know why I write this—bit jumbled, no doubt, but some numbers spoken.* | Sign on plane is N-214 |

## Test 22

### December 13, 1937.   11:30–12:00 P.M.

*Today I worked intensely until early evening on the synopsis of musical comedy. I had an 8:30 P.M. appointment with William Ortmann, composer, and his wife,*

Ilse Marvenga, at the Victoria Hotel to read them the script. There I met their business partner, Hubert J. Braun, of Chicago.

I discussed the play until just time to leave, and get home for my usual session.

Of late, I have noticed a little more interference of the conscious mind, since it has been some weeks without my having had any confirmation from Wilkins. Being able to check my impressions is always most helpful, and seems conducive to even better results.

Being able to distinguish between a "thought of action" which may be in Wilkins' mind—and the materialization of this "thought in action" is sometimes difficult.

An ultimate "break-down" of these impressions, and careful analysis may reveal much more of interest than at first-hand appears. In several instances, I have definitely had the feeling that I was "seeing" events before they had actually occurred. Only time, and a check-up of these impressions, can determine whether this has been so.

An exact transcription of tonight's impressions is as follows:

### SHERMAN

*Action tonight—seem to feel you in air—more snow since you left Point Barrow—bitter cold—has been 30 below or more—your face seems to loom up in consciousness—as though you looking straight at me . . .*

*Seems like something has to be soldered—came apart—due freezing—and broke off or cracked . . .*

*Trouble with one of skis—in rough landing—due conditions field and bad flying weather—forced down, either on return flight Aklavik or test flight—along coast line—or out over ocean ice—see great barren stretches underneath—occasional shack or outpost some sort—altitude 2,000 feet—seems to be level ordinarily maintained—Visibility hazy—air bumpy —considerable wireless activity . . .*

*Name "Allen" comes to me as though spoken—see fairly tall, slender man—*

### WILKINS

We made flight this day, but weather very cold—snowing—had to turn back. We were on our way to deliver gasoline to a plane which was forced down.

A brass pipe which we soldered—a flask of compressed oxygen which we were using to "blow up" (put air into shock absorber on tail skid) broke off, and we had to make one of a rubber tube, and bind it with wire.

We were forced back to Barrow as noted above.

Continuous wireless activity with and about the Fairbanks plane which was down.

Allen Dyne is our mechanic. We call him "Al." He is slender, but not tall.

| SHERMAN | WILKINS |
|---|---|
| *mechanical turn of mind—interested your plane—doing something about equipment—is such a man helping you? Get impression his working long hours—checking instruments, motors . . .* | He was naturally working hard on difficult job of getting air into tail skid. |
| *Your plane has to be pushed or pulled out of some spot—I see quite a group of men around it, as though helping . . .* | Groups of people around plane at take-off and landing. |
| *"Abandon search"—strange impression crosses mind—strong thought from some quarter—mind or minds consider further search futile.* | The "abandon search" was the abandon search for the Fairbanks plane. For when we got back to Barrow, there was a message for me saying not to come with the plane because a landing beside the other machine would have been dangerous to us in a heavy plane. |

## Test 23

### December 14, 1937.   11:30–12:00 P.M.

*With many things of a personal and business nature on my mind at present—more than ordinary—I am wondering if they have not impinged a bit upon my consciousness—retarding impressions from without.*

*A check on my work of the last three to four weeks would again prove a great stimulant. I find that a known fact is accepted by the consciousness, and is given no further thought. But the apparent facts I have telepathically picked up have a way of being challenged by my conscious mind as I check them over, without any means of actual verification at the moment.*

*Some day perhaps, man will discover how to bring his mental faculties under such directional control as to apprehend "facts" through extrasensory perception, and know them to be true, without any need for external confirmation.*

*I feel certain that when I have had more of this kind of experience behind me, my own ability to detect the false impressions from the true will have been greatly improved.*

*(A re-reading of my commentary relative to this night, will prove revealing, after a study of impressions below.*

*In addition to this uncertain mental attitude described above my first recorded thought this night was:*

"Sitting tonight—no immediate impression—usually several mental pictures or feeling impressions start to flow as though previously received by subconsciousness and waiting to be passed on."

*For scientific study, my having felt differently this night . . . as though I did not possess ability to make real contact, is worthy of analysis. All impressions this night seemed more related to Wilkins' thoughts than his actions—therefore incapable of "proof" or checking, as will be noted.)*
*An exact transcription of tonight's impressions is as follows:*

(Here, for first time in all my sittings, Wilkins has made no notations opposite my impressions for this night, most of which were more of a general nature.)

### SHERMAN
*(Impressions Unconfirmed by Wilkins)*

*Sitting tonight—no immediate impression—usually several mental pictures or feeling impressions start to flow as though previously received by subconscious—and waiting to be passed on. . . . "Back Aklavik" now seems come—preparation now for first search flight—weather permitting—You tonight, heavily clad—hood garment—black or very dark colored—with companions—very intent on something—serious time—I now seem to hear airplane motors—feel intense cold—see odd shadow of plane on ice cast by light of moon—weird sensation . . .*

*Another story* Times *very soon—feel you about send to Iversen—moonlight not enough for successful search flight—you must have fairly good visibility, and this may delay initial take-off . . .*

*Don't feel you've started real adventure yet—though have feeling of imminence—seems to be more a concentration of interests—a singleness of purpose—a waiting for right conditions with "more of a mind" to action than have heretofore picked up . . .*

*It seems as though most of details have been attended to—up to making of actual search flights . . .*

*Your mind centered more on your objectives now—I follow it out into Arctic wastes, and broad expanse of territory you hope to cover—You are giving careful thought to this tonight—unusual impression of your intentness—seem to see you in plane, watching progress of flight, and checking position—white light—all well—Wish I could be sure whether this tenseness is in me or reflected from you tonight—a strange feeling which seems to keep other impressions from coming through that I can vaguely feel on the threshold of consciousness—something appears to be happening that engages best attention your entire consciousness—Good night.*

## Test 24

### December 16, 1937.   11:30–12.00 P.M.

*This has been a week of intense creative work. Tonight I took several hours off to attend the lecture by Dr. Hardwicke at the Psychic Research Society, 71 West 23rd Street, and was home in time for this sitting.*

*I had been up practically all of the previous night, writing, and intended to remain up tonight, in order to finish a ten thousand word baseball story.*

*I mention this creative activity, so that it may be taken into account when the record of these impressions is studied.*

*In my own opinion, what came through tonight should have been more accurate, because my conscious mind had been so actively attuned to pulling material from the subconscious. When I relaxed tonight, it was with an indifferent and unconcerned attitude. I did not seem to be straining so much, or impelled by occasional doubts due to the fact that it has been a long time now since I have had the assurance which comes with a check on my impressions.*

*An exact transcription of tonight's impressions is as follows:*

| SHERMAN | WILKINS |
|---|---|
| Though this is day your announced planned take-off, can't feel you, this moment, in air—think condition today checks back to earlier, almost premonitory impression of weather preventing start search flight on December 16th . . . | Okay. |
| See you spending part of day reading over quantity of mail received by plane to Aklavik—and writing letters . . . | These messages and letters came from Fairbanks. |
| "Standing By" impression coupled with one of enforced delay—several things at last moment not right and storms—snow over route make visibility too bad for take-off . . . | Continuous bad weather. |
| Definite "grounded" sensations—and will be surprised if this impression is not exactly correct—temporary standstill, and nothing you can do about it . . . | Correct. |

## SHERMAN

*K seems to be smoking pipe—also yourself—as you sit in group talking tonight—informally—some other men present—seems like you're in front of a fireplace indoors—cheery, warm contrast to cold outside—a priest-like man comes in as though one of guests—another of military bearing—another—a pilot not connected your expedition—can he be mail pilot? . . .*

*Triangle impression—are you to fly out from a point or base which represents apex of triangle—then veer to right or left as case may be in angular line to furthermost point in Arctic—then swing across on straight line and come back to original base or point on angular line—thus giving yourself observational coverage this area? . . .*

*Nearest I can come to describing plan of flight which seems to be in your mind—I see some such lines drawn on a map—on next flight you appear to invert the triangle—going out along edge of territory previously covered to imaginary point in radius of Pole—then returning on an angular line, forming a triangle with its straight line base near your starting point, and rounding out a parallelogram as a result of two flights—of area covered . . .*

*It is difficult to put in words what I see or feel mentally with regard to way your flight is figured—but it strongly appears to be system of angles—which you hope scientifically to control and compute as you fly—so you can mark off various areas and chart them mathematically—this has given you much study . . .*

## WILKINS

The pilot who flew the plane from Fairbanks left Barrow that morning. We were anxious about him because of bad weather and he was forced down. We, a group of us, constantly on radio, listening for his report. He finally took off again, and reached Fairbanks.

Description of track on chart correct.

I am not sure that this was the day I marked the track on the chart, but I referred to it in a message to Fairbanks, and it was strongly in mind.

| SHERMAN | WILKINS |
|---|---|
| *Don't know why—but at different moments cross, or "X" has appeared. Have you received ESP cards from Samuel Emery?* | Cards not yet received. |

## Test 25

### December 20, 1937.  11:30–12:00 P.M.

*Despite the fact that it has now been some weeks since any word has been received from Wilkins, or since he has received any mail from me—a quantity of it no doubt now awaiting him at Aklavik—I am continuing to sit for impressions at the appointed times.*

*I am finding what I believed to be an increase in my powers of receptivity due to this regular schedule I have followed . . . and I feel sure that many interesting observations will be disclosed when a careful check can be made of these detailed impressions.*

*When Wilkins gets within the reach of direct radio communication, these tests should be greatly expedited. Also—when his work has become so routinized as to permit him completely to cooperate with me at times agreed upon.*

*Even so, if I have been successful in picking up many true impressions from his mind—and clairvoyantly—these results should be as significant as the tests we are still to perform.*

*I wish I could describe the unsettled mental feeling which has come over me the past ten days when I have thought suddenly of Wilkins, or during the periods I attempt to receive direct impressions. I feel certain this has been a difficult, disappointing time for him, with conditions beyond his control, bringing about further delay in search flights.*

*An exact transcription of tonight's impressions is as follows:*

| SHERMAN | WILKINS |
|---|---|
| *D seems to be a lucky winner at cards—I see him laughing about some triumph he has gained over certain members crew . . .* | Correct. We playing a card game at Radio station. |
| *Is there some dog or dogs members [of expedition] attached to? I seem to see them on ice and snow in vicinity of plane . . .* | Many dogs—one conspicuous St. Bernard. |
| *You have some rare wine offered yourself* | |

| SHERMAN | WILKINS |
|---|---|
| and crew tonight . . . I seem to see you all partake . . . | Blueberry wine—*not bad!* |
| "Might as well make the best of it," some man, important in life of Point Barrow, says. "We want to see you fellows accomplish your purpose, but we're also glad to have you stick around—and if you're here for the holidays, we want you to let us help you celebrate Christmas . . ." | As guests of Mr. Morgan, radio operator, most forceful man at Barrow. He warned me that night that I had, with the others of the crew, to have Christmas dinner at his house. One other time, when he expected me to come to dinner with the others, I already had an engagement. |

## Test 26

### December 21, 1937.   11:30–12:00 P.M.

*One of the difficulties in the receiving of thought forms is distinguishing between a thought in the mind of an individual, and the actual materialization of that thought in action.*

*I am certain that I have confused these two thought forms on a number of occasions over this period of time when I have recorded my impressions.*

*Since the mental picture in the mind of an individual of an external action is often only a little more vivid than a mental picture of a contemplated action—it is small wonder that confusion sometimes takes place.*

*Particularly do I believe this to have been true of my impressions of the past two weeks when I have felt distinctly Wilkins' intentions to take-off . . . his mind centered upon different objectives . . . and he has been thinking so strongly about these matters that I, in several instances, have been on the verge of recording an action which has not yet taken place.*

*I think it will be revealed that a number of things have been happening as a result of these experiments. With the mind made receptive, one's conscious definitions of time and space are at once altered . . . according to the degree of receptivity at any given moment. Not only is it then possible to perceive the past and present moment . . . but the future takes its place in the same pattern, and becomes coexistent for the split second that this contact can be maintained. In such a mental state, one is apt to "pull in" flashes of things-about-to-be, as well as things occurring in the present and the past.*

*I am trying to develop the ability to hold this inner faculty of mind for longer and longer periods of time—and to be able to determine whether the impressions received, in all instances, are related to past, present, or future.*

*Tonight's record of impressions is as follows:*

| SHERMAN | WILKINS |
|---|---|
| *Sudden severe pain comes to me—right side of head—I record this because it happened at this time and not, at the moment, because I attach any significance or relation to it—other than passing physical disturbance in me—sometimes, however, I seem to see or feel physical ailment affecting another . . .* | Am not sure that it happened this day, but each one of us could not seem to avoid bumping our heads on a sharp-edged stove pipe in the kitchen of our quarters. I bumped mine only twice but Dyne and Cheeseman bumped often. Cheeseman was "laid out" by the blow twice in one day. The pipe was at an awkward height. |
| *Work continuing on radio reception and sending . . .* | Very busy with radio. Word from Washington authorizing us to install some of it in the government radio house. |
| *Considering flight back to Aklavik in event not practical risk search flight soon—as Aklavik regarded more advantageous at present, of two bases . . .* | Correct. |
| *You have another project looming—to follow immediately after this work completed for Russian government. Think it in association Lincoln Ellsworth, and that further communications exchanged about it.* | Message from Ellsworth about his expedition to the South next season. |

## Test 27

### December 23, 1937.   11:30–12:00 P.M.

I have been unable this entire week to approach these appointed times without a "let-down" impression at the very start, as though things have not worked out right, weather conditions largely making search flights impossible.

I think this basic feeling has "blanketed" other impressions I might otherwise have received—but it has been so predominant, I am compelled to mention it for purposes of analysis later.

(Again I had a poor night for picking up impressions and so noted, as above.)

*(Only one impression was confirmed by Wilkins this night. Most impressions, as was case of night of December 14th, seemed more related to his thought processes than to his actions.)*

| SHERMAN | WILKINS |
|---|---|
| I seem to see Christmas mail and some packages stacked up Aklavik, feel eagerness men to go through it. | Correct. |

## Test 28

### December 27, 1937.   11:30–12:00 P.M.

Late Sunday afternoon, December 26th, I visited Reginald Iversen, New York Times *radio operator, in his radio room in the* Times *building for the purpose of relieving my mind.*

I had been so depressed for the past ten days, whenever giving thought to Wilkins, that I felt this state of consciousness was inhibiting to good telepathic results.

I showed Iversen my record of impressions for the past three weeks, and he said: "That's right, Wilkins is naturally disappointed at not having been able to take-off, and in having to postpone his search flights until the next time of the moon."

Iversen went on to say that Wilkins had been on the verge of taking off a number of times, only to have weather conditions prevent at the last moment, as I had recorded.

The little I learned from Iversen, in partial substantiation of impressions received, call for an analytical survey and comments which I shall later make, as time permits. I asked Iversen's reactions only to such portion of the impressions as would come within his field of knowledge. At no time have I asked him, nor has he volunteered knowledge concerning the nature of Wilkins' equipment or his activities. Fortunately, Iversen is a fair-minded skeptic; he appreciates the possible scientific value of the tests being made, and is cooperating excellently through giving out information only concerning a past recorded event which can have no bearing on future operations. The above statements, I am sure, he will verify.

It lifted a "weight from my mind" to know that conditions were much as I had mentally seen them, and that my depressed feelings were not induced by some state of being within myself. I hope that the following impressions received tonight, with a freer mind will show even more accurate results as a consequence.

*(Here is evidence that release of the conscious mind's assailment of doubt and wonderment as to correctness of impressions received immediately freed inner mental faculties for more effective operation.)*

| SHERMAN | WILKINS |
|---|---|
| *Christmas day you are entertained at dinner—yourself and crew—about eighteen present—was going to record fifteen at first, but seem to me three more added to party—several women present—serve you at table—room decorated, paper streamers, Christmas colors, no tree—something rigged up on table—candles, I believe—some important people Point Barrow join in—head of a trading post, a priest, a military man and his wife—you exchange some little gifts with some Eskimos . . .* | Correct. |
| *You are residing in some fairly large house—a stove in this house with long black stretch of pipe which goes up through ceiling . . .* | Correct. |
| *You have still had little chance to give concentrated thought my way at appointed times—though have thought of me often as you are tonight—Wondering what impressions I may have received during this long period of no possible contact or confirmation . . .* | Speaking of you and your work. |
| *Red flash tonight as though one of crew wasn't so well—hope this isn't true impression—it crossed consciousness quickly, but I record it in keeping with resolve to make known everything which comes to mind during these sessions . . .* | Cook, radio man at Barrow, sick. |
| *Yes, I see you talking to a doctor in Point Barrow with whom you have become quite well acquainted—he seems to be a thin, tall individual about fifty years of age . . .* | Dr. Levine of U. S. Government Medical Service. There for Research. Age about okay, thin but short. |
| *I seem to see plane partially covered with tarpaulin or canvas—new snow banking around it—some masts in air, like* | Men have been building new masts which augment the many there were at the radio station. These new masts quite close to plane. The new one of |

| SHERMAN | WILKINS |
|---|---|
| radio apparatus—building operation some sort—in vicinity of plane . . . | this style [Wilkins drew picture on original copy] slightly out of vertical. |
| You have watched some fishing through the ice—no, it seems like you have all tried your hands at it . . . | We men have not fished, but women (three) Eskimos, caught many small "Tom Cod," about eight inches long. They are very good eating. Did not record, but think the fish caught yesterday. An Eskimo woman was cleaning a frozen fish that night which she wrapped up for me as a Christmas present. |

## Test 29

### December 28, 1937.   11:30–12:00 P.M.

*Occasionally, during these sittings, there comes to me flash of an impending future event. Because this type of impression has no bearing upon the telepathic phase of these experiments, I have sometimes cleared my consciousness of these flashes. The skeptic might be inclined to consider such impressions as imaginative. However, I have had enough experience in the receiving of impressions of future events, which have later come to pass, that I have decided, henceforth, to record all such impressions here—for purposes of study and checking—whenever they shall occur.*

(It will be seen, by a study of this material, annotated by Wilkins, that I foresaw, this night, December 28th, the actual date, January 15th, that search flight would be made . . . the only date on which a search flight has been made or could be made [due to weather and other conditions] in the months Wilkins has been North!)

*An exact transcription of tonight's impressions is as follows:*

| SHERMAN | WILKINS |
|---|---|
| You have been on what amounts to a hike today—over barren country and ice and snow—you seem to experiment with some instruments at a distance from your base location at Point Barrow . . . | True. I was trying out a new pair of snowshoes. Went out over the pack ice and back over the land. |

## SHERMAN

*You wonder if Russian government situation stable enough to permit execution of ... promised submarine expedition aid—as reward for your search flights—you have already filed exhaustive report your submarine polar research plan with Russian government—which can be made basis of expedition of their own—did Russia so choose ...*

*These plans, also plans for expedition work with Lincoln Ellsworth—are in your mind in addition to details of present assignment ...*

JANUARY 15th *comes to me as day you actually make take-off for north regions—though you now hope to get off few days earlier in month—this again is premonitory impression—as though thoughts jumped ahead of present moment and this future moment, for a fleeting instant, becomes Now—then fades again—One may say this is imagination, but I named December 17th as day you planned first take-off several weeks ahead of actual time, and am advised you came nearest take-off on this date ...* (I had added in impressions this date: *"I cannot explain this type of impression or feeling ... except—when it comes* unsought—*usually quite positive in nature. Do not see weather conditions as favorable until then, coupled with time of moon."*)

## WILKINS

Heard about this time (think it must have been this night), from Iversen that Russians were studying my plan for a submarine expedition, but I had not received anything direct from them until today, February 6th, when a letter came from the Soviet Embassy with clipping of *Moscow News*, dated December 27th, headlined: "Arctic Expedition Under Sea Planned by Sir Hubert Wilkins, Closely Followed by Soviet Union ..." Some extracts were carried in New York papers the next day, I believe.

Ellsworth cut clippings about this (which I received today), and immediately wired again to be sure I would go with him.

*Quite extraordinary* that we did make the flight, starting night of 14th and returning on the 15th of January.

We should have gotten off on the 13th, but radios on plane had broken down.

## Test 30

### December 31, 1937.   11:30–12:00 P.M.

I attended a meeting of the Psychic Research Society this evening, and, as so often happens when I am in a place where many people are smoking returned home with a severe headache.

Despite this condition, I sat at the appointed time as usual, and it will be interesting eventually to determine whether the nature of my impressions was altered, as a consequence.

I might say, that after a half hour's relaxation, and placing my conscious mind at ease, I got up from the sitting much improved.

An exact "impression" record of tonight's sitting is as follows:

| SHERMAN | WILKINS |
|---|---|
| *I seem to see you walking along a snow-banked path—with a hill in the distance on your right—the path curves in this direction, and the hill, not very high, overlooks the ocean, now a mass, ice-caked and rough, and piled up in places along the shore or beach . . .* | Correct. |
| *Seems to be some New Year's dance or party to which you also invited—You and men in mood to welcome a little diversion to relieve monotony of waiting—I can almost hear you saying, "Happy New Year, Sherman!" as though sending me a greeting—knowing this is last time I'm to attempt to receive your thoughts this year—at any rate I send this mental greeting to you, and hope you get its impulse. Happy New Year and good night!* | Correct—native dance but I did not go. I went home to bed by myself, the others at the party—and thought of you particularly and others of my friends, saying "Happy New Year" to each one. |

# Test 31

### January 3, 1938.     11:30–12:00 P.M.

It becomes a bit difficult to maintain an alert, vital interest in attempting to receive telepathic impressions when a long time has elapsed between confirmations of material already recorded.

I think this is because the conscious mind, unrelated to this experience, persists in raising doubts and in questioning different impressions received. Once, however, the conscious mind is set at rest by an individual's knowing, one way or the other, as to the authenticity of certain impressions, the entire consciousness is relieved . . . and freed for further, more receptive experimentation.

*An exact transcription of tonight's impressions is as follows:*

| SHERMAN | WILKINS |
|---|---|
| *You seem to have had quite a time New Year's Day—visited around several homes—attended affair at night—had some liquor which was presented to self and crew—also cigars and tobacco . . .* | Liquor all gone. |
| *Presence of ladies helped evening occasion—Point Barrow's "finest"—a hospitable,* well-meaning *group—You in company woman in dark, I believe, black evening dress—seems to be wearing earrings . . .* | Correct. [Wilkins also wrote word "correct" underneath my descriptive word, "well-meaning," and wrote "yes" under "wearing earrings" . . . ] |
| *All manner of disturbing details and activities have come up at time you know I am sitting in New York—several times you have thought you could notify Iversen as to definite evenings you would try keep appointments, but have not quite been sure enough of being able to follow through.* | Radio communication with Iversen practically nil, and what little we could get through had to be sent twice or three times over. |

## Test 32

### January 4, 1938.　　12:10–12:41 A.M.

*Tonight, Clara Clemens Gabrilowitsch and her daughter, Nina, were out for a visit. They are very much interested in this subject, and they stayed until a few minutes after midnight.*

*This is the first time that I have not been sitting at the appointed time—but, since I have had no report from Wilkins that he has been able to keep these appointments, it seemed to me that I should be able to pick up as accurate impressions concerning his doings, even though late in sitting.*

*What came through tonight, as a consequence, should be of interest when it can be checked. I sat from 12:10 to 12:41 A.M.*

*The following is an exact transcription of my impressions:*

| SHERMAN | WILKINS |
|---|---|
| *Strong wind appears to be blowing . . .* | Some wind, but not much. |
| *You have course of action mapped out for self for next two to three years following completion present job—involves association Lincoln Ellsworth, your own submarine expedition, advisory and perhaps actual service further Russian exploration . . .* | Correct. |
| *You think of me, but still little time to carry on our work.* | Correct. |

## Test 33

### January 6, 1938.     11:30–12:00 P.M.

I am going to make an effort soon, through Reginald Iversen, New York Times radio operator, to reach Wilkins, and, in the event he has not received my mail awaiting him at Aklavik, to suggest some tests that he may institute, in line with Dr. Gardner Murphy's suggestions.

It will be helpful to me to have some sort of check on my impressions, if only of a limited nature.

The exact record of tonight's impressions is as follows:

| SHERMAN | WILKINS |
|---|---|
| *You much in company youngish looking man, a trifle shorter than you—appear to have much in common—he seems to have been in vicinity Point Barrow about two years—believe he has engineering background—interested in natural resources—particularly oil. You discuss matters with him . . .* | Mr. Morgan, radio operator. We often talk oil. There are oil seepages near here. He has invented a "distiller," which he wants the government to develop and furnish the Eskimos who may, by that means, recover oil. Now they buy oil, but can burn the black pitch from pitch holes nearby. |
| *Someone has entertained you with quite a collection of phonograph records—some of which recall old times.* | Correct.<br>A toy Christmas present machine with thin paper records—*old time tunes.* |

## Test 34

January, 9, 1938.

## Communication to Sir Hubert Wilkins at Point Barrow

*On this date, I recorded the following:*
*With Sir Hubert Wilkins still unable to secure his mail from Aklavik, giving instructions for cooperating along the lines suggested by Dr. Gardner Murphy, I finally moved today to make other, more immediate arrangements.*
*Through Reginald Iversen, radio operator of the* New York Times, *a radiogram was dispatched.*
*(This radiogram was never sent due to bad radio conditions . . . but my conscious mind so inhibited me as to "throw me entirely" my sitting night of January 10th, since I "consciously felt" radiogram had been sent. This demonstrates once more that the less a "sensitive" knows about a supposed condition, the more accurate his or her impressions. My conscious anticipation of Wilkins' receipt of my radiogram was so great as to have colored my inner thoughts, thus interfering with real telepathic reception. At this time, also, not having had a confirmation from Wilkins in weeks, I was battling my own conscious doubts and wonderments, feeling keenly too, the fact that Dr. Gardner Murphy possibly wasn't satisfied with my sittings, furnishing him with impressions which he could not check, as in the case of ESP card tests.)*

### SHERMAN

*(my proposed radiogram to Wilkins which was never sent)*
*(Copy of radiogram)*
Sir Hubert Wilkins,
Via courtesy of New York Times
Doctor Murphy of Columbia suggests quantitative tests, twelve tests nightly at previously agreed periods Mondays, Tuesdays, and Thursdays as follows: prepare cards numbering one to ten inclusive. Have one man shuffle entire pack each time and draw one card, handing you for mental transmission to me. Later radio via Times rotation numbers drawn. Will be glad to know if our previously arranged appointments are convenient to you. Kindest regards and best wishes.

### WILKINS

Have been unable for some time to get any messages through to or from Iversen, so naturally did not get this one. It may have come to Aklavik, but I doubt it, and I have not yet received it. Just possible Iversen might have sent it, and it mislaid at Aklavik.

Radio conditions have been even worse this year than when I first tried it with primitive machines in 1926.

## Test 35

### January 10, 1938.  11:30–12:00 P.M.

*It will be more gratifying to me the instant more definite arrangements can be completed with Sir Hubert Wilkins concerning our tests.*

*Inability of Sir Hubert to secure detailed instructions of these proposed tests, awaiting him at Aklavik, and his own duties concerned with the expedition, have prevented his conscious cooperation to the extent that he had planned.*

*My radiogram to him of January 9th may serve to clear up this difficulty, and produce tests affording an almost immediate check on accuracy of pick-up.*

*(As previously stated tonight's sitting a classic example of coloring which destroyed my ability to get any accurate impressions from or about Wilkins.)*

*An exact transcription of tonight's impressions is as follows:*

| SHERMAN | WILKINS |
|---|---|
| *You have received my radiogram through courtesy* New York Times *suggesting test—and so forth.* | See above. Did not receive radiogram. |
| *I have impression you have already made the numbered cards, from one to ten, inclusive, as suggested by radio . . .* | No mail or message yet. |

## Test 36

### January 11, 1938.  (After Midnight)

*Tonight was unusual in many respects. Earlier in the evening, I attended a lecture by Dr. Gardner Murphy at the Psychic Research Society, 71 West 23rd Street. Dr. Murphy's subject was:*

*"Some Current Methods Used in Studying Extrasensory Perceptions."*

*My report associated with my impression of that night, reads as follows:*

> "Meeting Dr. Murphy after his lecture, he made it clear to me, in the matter of my telepathic experiments with Wilkins, that he did not wish to inhibit either one of us by too specific suggestions. He has found, he stated, that different individuals employ different methods to obtain evidential results."

*Now observe this unusual attendant circumstance in the light of the telepathic results achieved this night, as I went on to report:*

"Returning home late, I was on the subway at 11:30 P.M., and started making notes of my impressions. I scribbled further notes as I hurried home down 10th Street and completed my work in my study, as hereinafter recorded."

*My talk with Dr. Murphy this time was the agency which provided a release of "conscious mind tension" . . . with accuracy of telepathic results ensuing.*

*Following are the impressions received on subway train enroute home from downtown . . . and scribbled as I ran along 10th Street to apartment house:*

| SHERMAN | WILKINS |
|---|---|
| *639 comes to me—don't know whether this is supposed to be time Point Barrow—6:39—or a number pick-up . . .* | 6:30 was time I set for myself to get up and prepare for flight to Aklavik. |
| *Hear motor turning over distinctly as though tested out—see you and crew running across snow-covered field in a hurry—concerned with some operation—activity of some sort tonight . . .* | We flew back to Aklavik this day. |
| *Lights on field—men gathered—you might decide to take off later tonight after all . . .* | A photographer with a long burning flare, taking "movies" of us as we got ready to leave. |
| *(11:48 P.M. Balance of sitting being completed in study at home.)* | |
| *Late weather report more favorable—Intent attitude of men tonight—feel them thinking and acting as a group—many people moving about—unusual activity for Point Barrow—Eskimo men and women, heavily garbed out watching . . .* | ALL CORRECT. |
| *Your mind now mainly on flight objectives—great determination to get off and complete first survey successfully—one motor acting up, due to cold—takes long warm-up—extra care on this—* | I felt this possibility, and took especial precaution for immediate help in case such did happen. |

| SHERMAN | WILKINS |
|---|---|
| *otherwise feel that motors do not possess proper lifting power with possibility of accident at take-off...* | |
| *Seem to see plane move along prepared runway on skis and then—with men helping—return to starting point after some difficulty—motors turning over for taxiing purpose—this appears to be occasion when men run along across ice and snow...* | Plane started off, then got stuck. We had to get out and jack it up, put wood blocks under the skis for the next start. |
| *Checking plane's traction and run-way conditions—fixing certain uneven or squashy spots—packing down and filling in—definite action preparatory to take-off, being carried on.* | CORRECT. |

## Test 37

### January 13, 1938.   11:30–12:00 P.M.

I worked intensely all day on a story for Boys' Life, entitled, "The Out-Boarding Boarder-Outer" a yarn about a boy, an outboard motor, an enormous appetite, and a fishing experience.

During moments in my creative work, I seemed to get flashes of Wilkins—impressions of disappointment and further delay. I stopped work on the story at 11:30 P.M., and took my seat at my other desk in the study, switching out the lights, and clearing my mind to receive impressions.

I mention the above as an illustration of the line of demarcation between the imaginative faculties of mind and the receptive, intuitive, or telepathic faculties.

An exact transcription of my impressions is as follows:

| SHERMAN | WILKINS |
|---|---|
| *Throughout the day, unbidden, has come the flashing impression that weather conditions again are causing postponement of flight intentions as you realize no purpose in making hazardous flight where visibility too poor to observe* | Weather was not good, but better than any during previous moon. In fact, would have taken-off except that both radio sets in the plane were "out of order," due to some delicate parts being affected by cold. |

## SHERMAN

*evidence of lost Russian flyers, were you over them—this disappointment intense and equalled only by intense desire to be off...*

*Trouble also with wireless sending—conditions none too good here, either—disturbance of one kind and another all along the line—am again impressed with 15th of January as your probable take-off date—that conditions won't permit attempt until then—despite your present standing by attitude...*

*You either have cards which you have made in anticipation these tests or message from me in your left hand coat pocket together with copy of scribbled reply—I see you take this something or several things out and look at them, thoughts on me at time...*

*Something of great importance mechanically has been found impractical and has to be changed—I get this feeling strongly and see you supervising work in connection therewith...*

*You really need a new or different device which you don't have with you—something is ingeniously made at Point Barrow by your mechanic in association with your radio engineer...*

*You seem to have talked about Australia to someone today—and told of some your experiences there...*

*I feel tenseness of crew due to uncertainty—a worse kind of suspense than actually in air.*

## WILKINS

Short wave wireless almost impossible. Have had to send nearly all messages—all, in fact, if I remember right, through government long wave channels.

Have had my message for you ever since reached Aklavik and received your mail. Received that on night of January 11th. (See mail in other envelope).

Radio.
Also means of cleaning plane's windows of hoar frost—working on it intently.

Correct.

Correct.

## Test 38

January 17, 1938.    11:30–12:00 P.M.

*I have had some very interesting confirmation of a number of important impressions through checking with Reginald Iversen,* New York Times *radio operator, following Wilkins' first search flight on January 15th as foreseen by me. As soon as time affords, I will prepare an analysis for study purposes.*

*One very definite point, however, I wish to mention here, illustrative of the care which must be taken not to suggest a line of action to a "sensitive."*

*Being eager for more definite results which could be checked by Dr. Gardner Murphy I prepared the radiogram with Iversen, which he said he would send on to Wilkins. He was, however, prevented from so doing . . . but my conscious mind, each sitting thereafter, was plagued with this wonderment . . . and tried to get me to ascertain Wilkins' reaction. Coloring naturally resulted. As a matter of fact, that radiogram has not been sent . . . and now that Wilkins has finally received his mail, with instructions from me, it probably never will be sent.*

*I will comment more about this in my analytical report when it can be made.*

*(This night I had learned for first time that Iversen had not been able to get radiogram message through to Wilkins, and so recorded in my comment as preface to listing of impressions. Wilkins made the following annotation opposite this information.)*

*"See now that the radiogram was not sent. I am reading these dispatches in order of postage dates . . . and checking against my diary."*

*An exact transcription of last night's impressions is as follows:*

| SHERMAN | WILKINS |
|---|---|
| *Now, of course, I know of your one search flight, made* January 15th, *as I had received the impression some weeks before, which persisted throughout—I have read the account of this flight in the* New York Times *of this morning's date, and am mentally relieved to have made this notation since it seems to clear my mind for any impressions I may receive from you tonight . . .* | Correct. |
| *Though you have overhauled your plane following the search flight preparatory for another take-off, having today in mind—weather conditions, clouds, and* | Correct. |

| SHERMAN | WILKINS |
|---|---|
| snow along route or course, have prevented further flying—You seem to be thinking strongly of another attempt tomorrow, if conditions change . . . | |
| I see you examining, with great interest volume of mail and many recordings of my impressions, referring at times to your diary—and making notes . . . | This done on 11th. |
| Tonight, I seem to be down on river ice with you at your radio or wireless tent—as though you trying to get communication through to Iversen, New York—several matters you wish to get across to him—conditions permitting—points necessary to clear up from correspondence you have studied . . . | Correct—always with message for you on file, which was not sent until February 3rd. |
| When you landed on your return, I seem to see mental picture of plane alighting on uneven surface and tilting to my right, on right side, as I would be sitting in the plane—with skis on this side not evenly supported by snow and ice beneath—I record this impression because it was not reported by you as flight incident—and I seem to see Aklavik or landing spot circled carefully, 2, 3, 4 times—before setting plane down—with men on ground waving arms and helping directionally—as well as through wireless—you seem to jack side of plane up and taxi it to different location after landing . . . | Correct.<br><br>We jack each side of the plane up after every flight, and put pieces of wood under the skis. |

## Test 39

January 18, 1938.    11:30–12:00 P.M.

*I returned home from downtown just in time to step in my study, and get seated at my desk by 11:30 P.M.*

*Whether right or wrong, I had been getting impressions, while listening to a lecture at the Psychic Research Society, by Dr. A. E. Strath-Gordon, that Iversen and Wilkins had been trying to communicate by radio.*

*I was disturbed by this impression since my conscious mind tried to tell me that Iversen was off duty on Tuesdays and Wednesdays . . . and did not come in to the* Times *office. However, this impression seemed to persist.*

*An exact transcription of what seemed to come through tonight is as follows:*

| SHERMAN | WILKINS |
|---|---|
| *Though this is supposed to be Iversen's day off at* Times, *have felt strongly, even before time for sitting, that he had been in radio touch with you today—that you have been communicating . . .* | [Wilkins made no comment opposite this rather unusual impression . . . But Iversen confirmed the fact that he had gone in to *Times* office today, on one of his off days, in hopes he could make radio contact with Wilkins. Reception still bad.] |
| | Sent word *this* night to that effect to Behakon at Fairbanks. |
| *Possibility now of having to wait until next month for further flights—Seem to hear you as though you spoke my name, "Sherman," mentally, if not aloud. You have shown some of my impressions to a few people and discussed them . . .* | Thinking of you each night when trying to raise Iversen on wireless.

Only one man here interested—Wilson. He "practiced" a little thought-form sending for some years, but, like myself, has been too much interrupted lately to do anything with it. |
| *Mind goes ahead to March 18th as significant, eventful day for you—I record this now—with wonderment as to what it can mean—you disappointed things did not work out as strenuously planned earlier today.* | Goodness knows what will happen on the 18th, now that our engine has gone. |

# Test 40

January 20, 1938.

*Tonight I attended a forum meeting of the Psychic Research Society at 71 West 23rd Street, and there again met Dr. A. E. Strath-Gordon whom I have known for some years.*

It was necessary that I discuss matters of importance with him, and since Sir Hubert has not yet confirmed his keeping of appointments with me, I permitted myself to stay out beyond the appointed time, as a further experiment.

While I seemed to feel his thought of me between 11:30 to 12 midnight, while seated with Dr. and Mrs. Strath-Gordon and my wife, at Childs Restaurant, Fifth Avenue and Twenty-fifth Street, I felt that I could pick up impressions a little later, since there was nothing that would have to be synchronized on an exact time basis.

Consequently, it was ten minutes to one o'clock tonight before I sat down in my study and made my mind receptive to outside impressions.

The following is an exact transcription:

| SHERMAN | WILKINS |
|---|---|
| *You seem to be with group of 6 to 8 people—they are laughing, talking, some drinking—little social occasion some sort . . .* | Correct. An "art" show at the Church of England mission station. Drinking tea, playing ping-pong, deck tennis, and so forth. |
| *One man, tall and slender, in earnest conversation with you, seems to have "S" in his name, perhaps one of his initials—You thought of me during period tonight—have played around with ESP cards—may be coloring, but get impression you've tried them out as a game with some of crew and Aklavik friends . . .* | Correct. The Rev. Sheppard is in charge of the Church of England school. We played games together. We talked of you and I tried "playing around with ESP cards" before going to the party, but did not concentrate since we had no definite arrangement. No one else except myself at Aklavik has seen the cards, although I have talked about them to Wilson. |
| *Will be surprised if I do not receive message from you soon, as I definitely feel you have framed one for me—and have made considerable notes on material sent you of interest to me.* | Same old message, still waiting, practically same as eventually sent. |

## Test 41

### January 24, 1938.     11:30–12:00 P.M.

*And still, due to conditions beyond control, it has been impossible for Wilkins to arrange more definite tests along the lines suggested by Dr. Gardner Murphy of Columbia.*

# Wilkins–Sherman Telepathic Tests

I have continually felt that Wilkins has the making of such tests in mind, and still has every intention of getting them accomplished before he concludes his expedition.

Meanwhile, I am keeping my end of the appointments, and recording what impressions seem to come.

The following is an exact transcription of tonight's impressions:

| SHERMAN | WILKINS |
|---|---|
| You have thought considerably about magnetic storms caused by sun spots and their effect upon short wave reception . . . | Correct—discussing this with Wilson. |
| It seems to be clear tonight for a change—or clearer than it has been—but too late now for you to undertake search flight until next month . . . | Correct. Diary says: "Sunshine today, bright tonight, 40 below." |
| You observe heavens and contemplate mystery of space—effect of heavenly bodies upon our earth—Are you looking at sky and thinking of me? I seem to get this impression. Thoughts go to stars in sky and more definite sensing your conscious thought my direction . . . | Correct. Thinking of you at radio, as in my room at appointed time. |
| You are out somewhere—I see smoke curling up from fire—three tents appear to be nearby—attempt at radio communication—still not good but perhaps a bit better . . . | Correct. Wood stove in radio tent, always smoking or steaming in the low temperature. Two tents. |
| You have message prepared for me when it is possible to get it through to Iversen—have had it for some time . . . | Correct. |
| Seems to me that some mail has left Aklavik for the States—appear to see plane activity, not connected your own plane, you appear to talk with mail pilot, remark on hazards winter flying and poor weather conditions—He has not been able to follow his own flight itinerary on schedule—and expresses himself just as well satisfied Aklavik his last stop. | Not since the 11th, then unfortunately before I had time to finish your notes. [Wilkins, according to Iversen, managed to get off hurried note to Lady Wilkins to catch this plane, at Aklavik days off schedule. I had no way of knowing this fact, however, but recorded earlier that Wilkins had received my mail.] |

## Test 42

### January 25, 1938.    (After Midnight)

*Once more I was detained downtown beyond my appointed time. This was occasioned by a conference with William Ortmann of the Braun Music Publishing Company, 50th and Broadway, who is doing the score for my musical comedy,* The Love Expert, *that is to star Ilse Marvenga.*

*Miss Marvenga sang the song numbers to my wife and me, and we discussed different story and music problems . . . with Mr. Ortmann.*

*But, shortly after eleven P.M., I got strong feelings of Wilkins, which continued intermittently until after twelve.*

*It seemed that he was trying to impress me with something unusual . . . and flashes of red would come to me while I was listening to the music.*

*When I got home, and started sitting at 12:30, I did not seem to bring through anything extraordinary . . . although I had the sensation that Wilkins had been thinking strongly of me.*

*An exact transcription of my impressions is as follows:*

| SHERMAN | WILKINS |
|---|---|
| *You seem to be doing considerable writing tonight—with notes and papers and letters about—as though you trying to catch up on correspondence, and other matters now that no flights can be made for some days . . .* | |
| *Only bad radio communication conditions are preventing definite plan of action on tests being sent to me—I am sure you have given quite some thought to this as you have gone over my material . . .* | Correct.<br><br>Yes. |
| *I can pick up impressions of you and your activities as I have been doing with considerable degree of accuracy—because I feel these impressions of what has happened to you are now in atmosphere—Specific impressions, such as deliberate conscious sending of numbers or symbols, seems to require a conscious knowing of* | *This point interests me,* up to the present, and has done for years. Ever since I conceived, somehow, of the idea of thoughts as concrete things.<br><br>Thoughts go on and on. The *strong* ones, "concentrated," that is, to the absolute exclusion of anything else in the mind or within the vision of the |

| SHERMAN | WILKINS |
|---|---|
| *transmissions at a given time, however, for effective results—I feel impelled to note this here, as though in answer to your own wonderment as to my ability to pick up, not alone diversity of impressions concerning your personal and expedition happenings but also numbers and symbols sendings . . .* | eye, by others. It is such thoughts that seem to be able to be "sensed" you have been picking up. |
| *You have certainly tried out ESP cards in some random moments and a rotation comes to me:* | |
| *[here were listed 25 symbols by me, but Wilkins reported he had not experimented by himself in this way as yet. However, this listing should be kept in mind for further check.]* | Not tried out in that way as yet. |
| *I seem to see picture you have taken of your bases and plane, and of members of crew which are to be used later by* New York Times *in possible rotogravure section—and otherwise.* | Was taking pictures *today* of plane, Aklavik, and so forth, with that strongly in mind.<br>Clear sunshine today. |

# Test 43

## January 27, 1938.    (After Midnight)

*Again I have been late in getting home for my appointment with Wilkins. But I have had the strange feeling on these occasions that many impressions have been picked up by my subconscious mind, and many can be pulled through by me the moment I make contact with it—through blanking the conscious mind.*

*It will be interesting to note, when a check can be made, whether my degree of accuracy has been apparently lessened on the nights when I have not been able to synchronize my time with the period Wilkins has agreed to attempt to send me telepathic impressions.*

*Always at 11:30 P.M. on the nights of the appointments, wherever I am, I commence getting strong feelings of Wilkins. Unless I am where I may clear my mind, I do not try to interpret these feelings, having found that to do this partially is like* exposing a film. It throws light upon the whole development and colors it with my imagination or conscious wonderments before I am ready to complete the entire operation. But, so long as I keep the impressions in my "dark room" until prepared to bring them out and "print" them through actual recording, I am apparently able to retain them.

An exact transcription of tonight's impressions is as follows:

| SHERMAN | WILKINS |
|---|---|
| Strange as it seems to record, you appear to be dancing tonight—or listening to dance music by radio . . . | Played victrola. Some tango records tonight, *first* time since arrival in north. Also trying to learn Russian by linguaphone, but alone as usual. |

*[Note the two following impressions . . . one later proven to have been premonitory, tying in with another "preview" impression that mechanical mishap would again delay search flight . . . The second impression hereinafter listed and confirmed by Wilkins, relating to unusual pick-up of personal happening.]*

| SHERMAN | WILKINS |
|---|---|
| Crank case of plane comes to me suddenly—did something go wrong with it, due to cold? . . . | Wonder if this was "forethought" or "preview knowledge" since it was not until February 6th that *we developed serious trouble in crankcase*—main bearing of one engine ground to powder that day. Must have been some trouble there since January 15th. |

*[Note connection January 15th with above happening . . . and my strong premonitory thought of January 15th as day of actual search flight—this day also apparently had even more significance as it was related to Wilkins.]*

| SHERMAN | WILKINS |
|---|---|
| A dog seems to have been injured in Aklavik and had to be shot—was injury sustained in fight with others—or something falling on it? Quite a strong feeling here. | Out walking—came upon dog dead on ice—it had been shot through the head—thought about it strongly for some time, wondered reason for killing. |

# Wilkins–Sherman Telepathic Tests

## Test 44

### January 31, 1938.    11:30–12:00 P.M.

*I returned home tonight from a conference with William Ortmann, who is writing the score for my musical comedy, just in time to keep my appointment.*

*Having seen Reginald Iversen,* New York Times *radio operator, on Sunday evening, I knew he was to attempt to communicate with Wilkins tonight, but, on the way home, had an unsettled feeling about it.*

*Upon arrival, my wife informed me that Iversen had phoned, saying air conditions were so bad he had been unable to pick up Wilkins at all.*

*I learned, upon yesterday's visit with Iversen, that more impressions of mine, received recently, are apparently very accurate . . . but I am so pressed for time at present that I am unable to record the many evidences of confirmation which now exist.*

*(Material returned with Wilkins' letter of February 10th, received by me, February 16th.)*

*An exact transcription of tonight's impressions is as follows:*

| SHERMAN | WILKINS |
|---|---|
| *Never before, in all your exploration and flying experience have you encountered such bad weather and magnetic conditions impeding not only to your search flight activities but to communication channels—constant delay and disappointment have attended you throughout on this score—and have a feeling that this is destined to continue . . .* | Correct. |
| *Seem to hear you humorously say to someone, "If radio communication keeps on this bad, I'll have to develop telepathic communication with Sherman in order to get anything through!" I have felt your thought my direction many times, and feel it again tonight.* | Correct. |

## Test 45

### February 1, 1938.    11:30–12:00 P.M.

*I finished tonight, after an intensive three weeks' work, the musical comedy book of* The Love Expert. *My mind has been filled with this creative effort almost continuously, save for the half-hour periods when I have tried to blank it for the purpose of receiving telepathic or clairvoyant impressions.*

*I make this notation for the purpose of later analysis. It will be interesting to note if the accuracy of impressions has been sustained throughout.*

*An exact transcription of tonight's impressions is as follows:*

*[It must be remembered that when these impressions were recorded, I still had had no confirmation from Wilkins . . . and the necessity of my combating a growing tensity of body and mind, as a result of my conscious mind's wonderment, was growing daily.]*

| SHERMAN | WILKINS |
|---|---|
| Had flash of Iversen, earlier this evening, with headphone to his ears as though trying to pick up your radio signals—but got impression communication still very difficult . . . | Correct. |
| Plane service irregular and not on schedule . . . | Heard that plane would leave ahead of schedule. Gave it considerable thought. |
| Need for certain kind of tubing crosses mind . . . | I was looking for a piece of steel tubing from which to make a ski scraper. Had a job to find it, and consequently thought "hard" about it. |

## Test 46

### February 3, 1938.    12:45–1:30 A.M.

*Detained again downtown, and unable to sit until above time. Was at Psychic Research Society listening to talks by Dr. Hardwicke and Dr. Strath-Gordon.*

*The following is an exact transcription of tonight's impressions:*

| SHERMAN | WILKINS |
|---|---|
| *Plane expected soon from Edmonton which will bring mail and supplies and enable you to get more mail out—Feel contact has been irregular in mail service . . .* | Expected Saturday, 5th. It arrived on the 6th. Mail received *only twice* since we reached Aklavik—at Christmas and *Feb. 6th.* |
| *Your next flight seems to be figured roughly as consuming between 1,000 and 1,200 miles out and back—conditions permitting—with alternative in your 900 mile flight—if considered advisable to turn back before complete coverage is made . . .* | 1265 out, and 1265 back. |
| *You have charted course on map of proposed search flight—and have gone over plans carefully with crew . . .* | Correct. |
| *I seem to see five ESP cards laid out on a table—face up—a cross, a star, a circle, a wave, and a square (left to right), and sensation of your trying to match them.* | Later (Friday) 4th, went over cards as follows:<br>Star, cross, circle, square, wave. |

## Wilkins–Sherman Special "ESP" Test

February 4, 1938.   11:30 P.M.

*Tonight, at nine o'clock, Reginald Iversen phoned me to say he had at last been able to contact Wilkins by radio and to arrange for a definite test along lines suggested by Dr. Gardner Murphy of Columbia. The first time was set for tonight at 11:30 P.M.*

*Wilkins' message to me was:* "Will try cards tonight and again after return from flight. Five rotation, each one minute."

*There is a time difference of about three hours between Aklavik and New York City, with Wilkins sitting at about 8:30 P.M.*

*Since this was the first time I had actual knowledge of Wilkins' working with me on a special test, my job was to keep from getting inhibited, if possible, by such a realization.*

*Only those who have had long and deep experience in this work can appreciate the task of keeping one's mind free; of completely stilling the imaginative faculty; of preventing a forcing element from breaking in; of avoiding becoming over-anxious.*

*Picking up definite impressions such as symbols thought of at a definite time is, as I have stated before, a more difficult achievement than receiving mental picture impressions of events concerning an individual. This is because one's conscious mind has no control over these events . . . but, in the case of trying to receive impressions of numbers and symbols, the conscious mind can attempt to tell the "sensitive" what the numbers and symbols might have been—which were not received.*

*I have been only too willing to cooperate with Dr. Gardner Murphy in the effort to see what can be done, at this great distance, along this line—but I feel that I should emphasize the importance of widening the investigative field to include more than a statistical checking of number and symbol impressions, either clairvoyantly or telepathically received.*

*The following is an exact transcription of my sitting tonight:*

| SHERMAN | WILKINS |
|---|---|
|  | Think the general pick-up as good, if not better, for our tests. |
| 1. Circle | Star |
| 2. Star | Cross |
| 3. Cross | Circle |
| 4. Circle | Square |
| 5. Wave | Wave |
| *You dress for outdoors—in fact you already appear to have been heavily dressed during time of experiments—and now go outside again—some work still to do—Good night.* | Correct.<br><br>Had to go to Radio tent. |

*[Note: This was last recorded impression of special ESP test, and indicates, despite coloring when test off mind, I made contact with Wilkins.]*

## Test 47

### February 7, 1938.     11:30–12:00 P.M.

*Samuel Emery, one of the witnesses to these experiments, reported to me that he had visited Reginald Iversen, at the* New York Times *Radio Station on Sunday afternoon, and found Iversen still skeptical but interested and willing to cooperate to the best of his ability in the relaying of advices between Wilkins and myself.*

*Iversen had received word from Wilkins that he would conduct the first ESP test last Friday night. I, accordingly, sat for impressions, and mailed a copy to Samuel Emery, Dr. Gardner Murphy, and Iversen.*

*Iversen is now awaiting a message from Wilkins giving the rotation of the ESP cards selected, so that this may be checked with the rotation I seemed to receive.*

*Radio communication, unfortunately, has been so extremely bad on short wave for some months, that there can be no guarantee of how often or when contact can be made with Wilkins.*

*I can say now, without having gone any further, that this continued concentration at regular periods has seemed to open up new inner channels of mind; has awakened an unusual dream area which would be interesting to report to those scientifically interested. Only pressure of other activities has kept me from keeping a record of these remarkable dreams or vivid impressions coming to me as I am about to drop off to sleep, following a period of concentration—which appears to deal with other civilizations, at times, and then again to be associated with remote parts of this or other countries.*

*I feel that there must be some factor in my period of concentration—willing my mind toward Wilkins—which is preparing me to pick up impressions of this kind, for they have never happened before. In fact, I have seldom dreamed in my whole life . . . although it does happen that I have dreamed, a few weeks in advance, of the demise of different members of our family—every one of these dreams being definite, and capable of verification: having to do with the loss of my brother, Edward; my father, and my grandmother.*

*These dreams have had no associative connection with any such suggestive fact as the illness of any of these relatives. When the dreams took place, they were in excellent health.*

*A recounting of these dreams has no point or purpose here, and would require too detailed a telling—remarkable though they might seem. I merely mention them because they are provable, and because they deal with a future-time factor phenomenon which science does not yet admit can be contacted.*

*An exact transcription of last night's impressions is as follows:*

| SHERMAN | WILKINS |
|---|---|
| *(Part of comment as preface to recorded impressions, this night)* | |
| *Reaching Wilkins by mail or getting mail out to the civilized world has been almost as difficult—so these tests, from the specific standpoint, have not worked out as desired.* | |
| *I have, however, despite all conditions, kept strictly to schedule. Only an exhaustive* | A fine show of interest and determination. Wish I could throw off other |

| SHERMAN | WILKINS |
|---|---|
| *check of all material compiled when all factors are known, can yield a proper evaluation of what has been received.* | uninteresting engagements, and keep schedule more often. |
| *So far as I know, you have not yet been able to radio Iversen rotation of five selected ESP cards which you attempted to send me telepathically last Friday night . . .* | |
| *Neither have you advised me, through Iversen, that you have set aside any further period of time for a repetition of this test—and so, it may be coloring when I record that I feel you are thinking of attempting another test tonight at my regular time—trying to determine if you can so impress me without notification.* | *Correct.* Same order as listed in other letter. |
| *cross—square—star—circle—square comes to me—not with definiteness former sitting.* | [This unusual *pick-up* indicating actual telepathic contact, definite time.] |
| *I could not feel you in flight all day Saturday, and sensed that Gillam had gotten through to Point Barrow without need of assistance from you. As for you— you were not anxious to make flight, and just as glad it proved unnecessary . . .* | *Correct.*<br>Gillam had radioed saying a landing with our plane would be extremely hazardous. |
| *I seem to see the moon tonight—not full enough for you to make flight—but visible to you—it lights up ice and snow in vague outline—I seem to be near point of land, and dark, almost black jut of rock—with dark spots amid white in distance . . .* | *Correct.* |
| *I see the tail of the plane—something done about the rudder control, it seems—to prevent its sticking in extra cold weather—I see radio activity—a hand tapping a key—sending messages—trying to make contact . . .* | On Tuesday, morning of the 6th, I thoroughly inspected the rudder control with the idea of using it for making up a sled should we "crack up" on the ice. Gave it some deep thought. This on the 5th. I *personally* sending |

|  SHERMAN  |  WILKINS  |
|---|---|
|  | messages to Barrow for the *first* time. Much concentration. |
| *Something fixed today with aid of a ladder of some sort—see one of crew reaching up . . .* | This on the 6th. We using ladder to put house over the damaged engine. Was risky job. Gave it heavy concentration. |
| *A plane taking mail from Aklavik—which means I shall be receiving mail from you soon . . .* | *I was on the plane.* Stayed at Fort Simpson that night. |
| *Seem to see little colony of tents on ice—three of them—the bigger in front—and other two behind—almost side by side, with space between—radio poles or mast above.* | *Three* tents—but scattered in triangle about 100 yards apart. Radio masts and poles near two tents. |

## Test 48

## February 8, 1938. (Saturday at 1 A.M.)

I was detained downtown on matters relating to my musical comedy, and was unable to sit until 1 A.M.

I was mentally disturbed tonight at exactly eleven-thirty, and looked at my watch while in conference at the City Club. To my surprise, it was only ten-thirty. I did not find out until an hour later that my watch had stopped at ten-thirty, and it had actually been eleven-thirty at the time.

I would have been late home for my appointment in any event, but it is interesting to note that I had a strong thought of Wilkins exactly at our appointed time, even though the evidence seemed to indicate I was an hour early in this feeling.

Joseph Deutsch and Michael Weisberg, in whose company I was, can testify as to my mistaken idea of the time, since they insisted it was an hour later than my watch had shown when I looked at it—as they glanced at their wrist watches at the same time.

It is easily possible that my long concentration upon the 11:30 time should have caused my subconscious to make me conscious of this moment when it arrived, rather than a direct impression from Wilkins. It will be interesting to know, however, whether he really was thinking of me strongly this particular night at the appointed hour . . . since I had other important matters on my mind, was confused as to time, and the flash of Wilkins came through anyway.

*An exact transcription of tonight's impressions is as follows:*

| SHERMAN | WILKINS |
|---|---|
| *You appear to be busy at something outdoors tonight—under light from oil lamps or lanterns . . .* | Was at Fort McMurray outdoors considerable part of time fixing ski on plane. |
| *Get impression you were thinking of me strongly at 11:30 my time tonight, but was at City Club in conference and could not follow up on impression—although a star flashed in mind . . .* | Was in my room at Hotel and thinking of you at 11:30 your time, but had to go out as somebody called me. *[Could "star" have been pick-up of his first "ESP" selection, tardily received by me but undoubtedly having been given strongest thought by Wilkins?]* |
| *I see you talking to a short, stubby little woman beneath an arched doorway— She is looking up at you and smiling . . .* | As I was going out, was waylaid by one of such description in the doorway. She had a collection of autograph books for me to sign. |
| *Some geologists or commercially interested men have asked you to keep eyes open for estimate of natural resources in land you are traversing—areas of mineral or other deposits . . .* | Spent the evening, more or less, discussing as we worked geological and mineral conditions in Canada. |
| *You are trying to reach an accord with the Soviet Union, since hazards have proven what they are—with weather so bad, flights could not be made on schedule or anywhere near it.* | Wired them twice during day *re:* replacement of engine. |

## Test 49

### February 10, 1938.   11:30–12:00 P.M.

*I was home on schedule from an afternoon and evening's work on the musical comedy.*

*Dr. and Mrs. Strath-Gordon were due out to see us around eleven, but when they had not arrived by eleven-thirty, I went into my study to keep the appointment.*

*The Strath-Gordons came at eleven-forty. I vaguely heard the doorbell, was conscious they were coming in . . . then forgot about them until midnight. This consciousness one takes on seems to be an awareness of a distant point and one returns to a consciousness of immediate surroundings at intervals, caused by some unusual sound close at hand . . . or, under directional control of conscious mind.*

*An exact transcription of my impressions is as follows:*

*[I sat on schedule but had been in severe abdominal pain, with stomach upset, for two days. As I recorded, the Strath-Gordons called just as I had started sitting. Dr. Strath-Gordon had come in, after a lecture downtown, to give me a physical examination. My impressions, as a consequence, seemed to have been affected. I did, however, feel strongly a previously received premonitory impression—and repeated it.]*

| SHERMAN | WILKINS |
|---|---|
| *Feeling crowds in—impression that came to me several weeks ago—as though unexpected development—mechanical mishap this time perhaps—would compel still another delay . . .* | This impression fulfilled. The mishap really occurred, I think on January 14th. |

## Test 50

### February 14, 1938.   11:30–12:00 P.M.

*I have been tremendously heartened by receipt today, after these many weeks, of all my recorded impressions from Wilkins, dating from November 15, to and including January 27th.*

*Suffice it to say here my number of direct "hits," as reported by Wilkins, has been "remarkable." In reading his marginal notes, I was again and again amazed at the accuracy of my recorded impressions—feeling strangely as though I had had nothing to do with these phenomena, now that I was considering the whole matter with my conscious mind. It gave me the unusual feeling of an outsider considering the material.*

*There remains now a big job on my part to type these checked impressions and make an analysis of same, which I shall do as soon as time affords, including a summary of our first ESP test.*

*I will only quote here several lines from Wilkins' letter, having to do with the impressions received, as a whole.*

*"You seem to get all the very strong thoughts, and sense the vivid conditions even though I am generally unable to get down to concentration at the appointed time . . . the main point is, that in spite of a lot of coloring and leads by the conscious, you have a remarkable number of 'coincidences' impossible of 'reason' or fore-knowledge,*

*and that, while it may not serve the purpose we had in mind, is, I am sure, helping to develop your perceptibility."*

I had received a premonitory impression, recorded twice, that Wilkins would meet with a mechanical mishap on his next attempted search flight, which would again delay him. This actually happened a few days ago—trouble developing in crankcase of motor (also recorded by me in advance of happening and confirmed by Wilkins). One motor was severely damaged due to plane propeller striking stake in ground on attempted take-off, necessitating Wilkins taking mail plane to Edmonton to secure new motor, where he is now. These circumstances will be covered in detail in my analytical report later.

An exact transcription of tonight's impressions is as follows:

| SHERMAN | WILKINS |
|---|---|
| *Get impression you've talked three times before different interested groups since arrival in Edmonton—first time before some luncheon club like a Rotary Club—telling them something about your mishap and intended plans for looking for Russian fliers in new place—as if they were inland somewhere . . .* | Rotary Club. |
| *You have found a motor in Edmonton, and it seems that you are planning to take off with it tomorrow or Wednesday if weather permits—Word "Mackenzie" flashes to mind in connection with flying—is there a company of some sort that supplies you with plane? You seem to be planning to investigate some report on Russian flyers made by Eskimos—which may give you a line on their whereabouts . . .* | Wednesday. Mackenzie Airways plane brought me out, and was preparing to fly me back.<br><br>Correct. |
| *Believe you are writing me again from Edmonton—checking more recent material and getting back to me—feel you particularly impressed my premonitory feeling your mishap . . .* | Correct. |
| *You in communication Aklavik today—Kenyon seems to want more supplies or parts of equipment . . .* | Correct. |

| SHERMAN | WILKINS |
|---|---|
| *Army officer—man in military attire, confers with you—some arrangements made, a favor asked . . .* | Saturday, 12th. |
| *You have to delegate New York, through Iversen, to secure some pieces equipment and rush through in relation to repair job on plane's motor—also in connection radio . . .* | Radio for Kenyon. |
| *Seem to see you as guest of church brotherhood—Sunday occasion—called on to speak—you required to pursue business matters on Sunday as well—do some long distance telephoning—believe you talk to Lady Wilkins . . .* | Correct. Addressed Knox Church (Presbyterian) on Sunday night. Wrote business letters all day. Telephoned to Lady Wilkins, but forget which night . . . |

February 15, 1938.
Dr. Gardner Murphy,
Department of Psychology,
Columbia University,
New York City.

Dear Dr. Murphy:

It has occurred to me that you might like to meet Reginald Iversen, *New York Times* radio operator, and get acquainted with him personally, since he will be the key man between Wilkins and myself when further ESP tests are carried out.

I want to be certain you are entirely satisfied with the test conditions, so that my mind will be freed of all tension.

I misinterpreted Wilkins' radio message on the first test—'five cards selected in rotation'. . . thinking they were to be chosen from pack of twenty-five, and thus possibility of same card being repeated . . . which caused 'coloring' in own consciousness. I scored one hit out of the five, as follows:

*Wilkins:* Star—cross—circle—square—wave
*Sherman:* Circle—star—cross—circle—*wave*

If you feel the test would be more convincing in the future should I not notify Iversen of my impressions until after he has received his report from Wilkins, I should be glad to work it this way—and let you be the one to contact him. In this first test, Wilkins was unable to radio the result to Iversen, and notified me direct. He still does not know what impressions I received.

Iversen still thinks this is "pretty screwy business," to put it in his own words—but he is a grand fellow and will cooperate to the best of his ability.

Incidentally, if you will study the above line-up of impressions, you will note that, allowing for the first miss—I had the order perfect for a run of three—jumped the fourth—and got the fifth in order. Wilkins told me he kept repeating his concentration in this order for a full half-hour.

Please let me hear from you. Regards!

Sincerely—

## Test 51

### February 15, 1938.

*It was not until 1 A.M. tonight that I was able to sit, because of an evening spent with the Thomas Buchanans where I met the East Indian, M. H. H. Joachim, who is much interested in mental phenomena. The Strath-Gordons were also present.*

*The following is an exact transcription of what came through tonight:*

| SHERMAN | WILKINS |
|---|---|
| *This seems to have been day of intense activity for you preparatory to leaving Edmonton for Aklavik . . .* | Correct. |
| *You are arising very early Wednesday morning with flight in mind, I believe—I see plane in air—man piloting whose initial seems to be "R"—rather stocky build . . .* | "AR"chie McMullen, pilot. |
| *Large box-like, crated affair seems to contain motor you have brought to replace damaged engine . . .* | Engine in large square box. |
| *Good time—party—some of the boys back in Aklavik—women—one of crew—feeling a bit too good—is warned by someone—doesn't like idea of your finding out when you get back . . .* | Something of this sort went on, but don't know if this was the day. Heard of it when I returned. |
| *You send off several wires today—believe* | |

| SHERMAN | WILKINS |
|---|---|
| number dispatched to be five—two in reply to wires received by you . . . | Probably (?). |
| Believe another note to me is on way or has been written. Feel definitely you have noted down what amounts to short letter with me in mind . . . | Correct. |
| Someone laughingly presents you with good luck charm, saying it's "to change your luck"—something you put in pocket, I think—color black—seems have some connection. | Half remember Miss Shackleton giving me something, but don't know what. |
| Feel you anxious over eventual synchronization of two motors—old and new . . . | Correct. Because different shaped propeller. |

## Confirmation by Wilkins of Impressions Received Relating to Secretly Recorded Dream

February 17, 1938.

On February 14th, shortly after noon, I arrived at the City Club and encountered George Ford, a theatrical friend of mine, whom I met there occasionally. While talking to him, I saw the mailman come in, and, because I had had a previous dream impression (which I had sent in a sealed envelope to Dr. Gardner Murphy and Samuel Emery) that I would finally receive a big envelope from Wilkins at a time when I was talking to George Ford at the City Club, I excused myself and went to the desk. Sure enough—two large envelopes of mail were turned over to me from Wilkins! I naturally felt a strange sensation at this unusual confirmation of the first portion of the dream . . . and opened the envelopes, glancing at their contents. Then I went back to the lounge to locate George Ford, and to ask him to witness my just having received this mail from Wilkins while in conversation with him. I was told by Paul, the doorman, that George had gone out while I was at the desk.

Further to substantiate this happening, I went immediately upstairs to the room of Samuel Emery, and recounted my experience to him—telling him the portion that had been in the sealed envelope in his possession, with relation to George Ford. I then read the Wilkins material in his presence, and, when I had finished it, returned to the club lobby, envelopes in hand, to again run into George Ford.

I showed Mr. Ford the Wilkins envelopes—also the date of the letter from Wilkins—told him of the experimental work I was doing without going into any detail whatsoever, asking him if he would be kind enough, for my records, to write me a little note, testifying to the truth of the above circumstances herewith related. His note follows:

February 14, 1938.
My dear Harold—
This is to attest that I was with you at the club today when a bundle of mail came in from Sir Hubert Wilkins, and that it was dated in his hand, February 6th.

Yours very truly,
(Signed) George Ford

I now feel free to reveal the first part of this dream which occurred the morning of February 3, 1938, at 9 A.M. I do not wish the sealed envelope to be opened by Dr. Murphy or Mr. Emery at this time, since the rest of the dream has not been fulfilled . . . and I do not want any other minds to be anticipatory of the happening therein recounted, for fear I would get a mental pick-up of such a reaction which would bring about "coloring" in my future sittings.

The following is an exact transcription of the first part of the dream, as recorded in the sealed envelopes—that portion which has now been confirmed by the subsequent developments of the past few days.

## Dream Experience (Sealed Envelope)

This morning, after my wife had arisen to see our younger daughter, Marcia, off to school—I rolled over for a last little nap and had the following dream:

I seemed to have stepped into the City Club and found a heavy envelope, especially bound, from Wilkins, postmarked "February 6th," containing copies of my impressions covering more than two months, with his usual marginal notes; also two enclosures in longhand, seemingly letters from two different persons purporting to have psychic ability, written to Wilkins; also a two-page typewritten letter from him. (This last seemed surprising to me, for I had not thought of his having a typewriter in the far north.)

At the time I picked up the letter at the Club, I seemed to encounter George Ford, a member who was a theatrical producer, and who wanted to talk with me about something. I recall being annoyed, as I wished to get away and go over the material.

Wilkins' letter contained a report on some matters not referred to in news accounts—having a bearing on my impressions—and began with an apology for his having been harassed with a multitude of details—"always something coming up to interfere" . . . Unusual, unparalleled conditions in North this season also accountable for Wilkins not having the free time anticipated to give to the tests.

Specific information and comments in this interesting letter almost come back to me in my conscious state but cannot quite recall . . .

(*Further contents of sealed letter withheld for time being.*)

## Additional Points of Confirmation Concerning This Portion of Dream

1. The "heavy envelope, especially bound" proved to be two heavy envelopes, manila, extra flap or seal on back, big enough to contain my returned typewritten pages mailed flat.

2. The material returned, containing marginal notes made by Wilkins, dated from impressions received by me the night of November 15, 1937, to and including the night of January 27th, 1938— *"covering more than two months"*—as recorded by me in the dream.

3. The envelopes contained two letters, written in longhand . . . but they were both written by *Wilkins,* instead of coming from two different persons "purporting to have psychic ability, written to Wilkins," which had been my dream impression.

Then I had gone on to record that I had received a two page typewritten letter from him. (*This last,* quoting from the dream recording, *seemed surprising to me for I had not thought of his having a typewriter in the far north.*)

4. However, yesterday, February 16th, I received another communication from Wilkins—enclosing a *two-page typewritten letter* from a young man by the name of T. W. Deachman of Ottawa, who told of the telepathic attempts of his fiancée, Florence Smith, and himself, in relation to Wilkins.

Quite obviously, the dream impression of this *typewritten letter* became confused in my conscious mind as coming from Wilkins . . . I saw clearly the two letters in longhand, written by Wilkins . . . I distinctly recorded *"two enclosures . . . from two different persons purporting to have psychic ability, written to Wilkins"* . . . and this *one typewritten letter* tells of the experiences of *two different persons.*

5. The second of Wilkins' notes to me, received the day I was talking to George Ford at the club is *dated,* not postmarked, *"Aklavik—night of February 6th, 1938."* My dream impression had caused me to list the date as postmarked, but the mail plane had arrived that day at Aklavik, expecting to take off then, but it actually did not until the morning of February 7th. However, Wilkins' letter was written and dated February 6th.

6. It must be remembered that this dream impression was received the morning of February 3rd. The night of February 3rd, at my regular sitting for impressions, I recorded the following:

*Plane expected soon from Edmonton which will bring mail and supplies and enable you to get more mail out.*

This most recent communication from Wilkins, returning my recordings of this date, contains this marginal confirmation:

"*Plane expected Saturday, 5th—it arrived on the 6th.*"

I then had gone on to record:

*Feel that contact has been irregular in this mail service as it has been in radio communication—due general unusual conditions.*

Wilkins' comment concerning this impression follows, a further substantiation of my dream impression date of February 6th:

"*Mail received only twice since we reached Aklavik—at Christmas and February 6th.*"

On the night of February 7th, sitting again for impressions, I recorded the following:

*A plane taking mail from Aklavik—which means I shall be receiving mail from you soon . . .*

In the letter written to me "the night of February 6th," and received by me February 14th, Wilkins says:

> "Although I had about two hours sleep the night before last, and hardly any last night (having been carpentering, and building a shack over the broken-down engine all day), and it is now 4 A.M., I have to catch the mail plane for Edmonton at 8:30, I am determined to send you a few lines and put the notes and this scribble into an envelope, so as to be sure to get them posted as soon as I get to Edmonton . . ."

The above notation by Wilkins, without his now knowing it, is perfect confirmation of my impression of arrival of the mail plane on the 6th, and its take-off from Aklavik on the 7th.

I did not receive the impression of Wilkins leaving for Edmonton on the plane . . . and put this down to inhibiting by my conscious mind, despite the fact I had received premonitory impression of "mechanical mishap and crankcase trouble," all of which has since been confirmed . . . But the willing of my conscious mind to Aklavik, where I knew Wilkins to be, apparently gave me such a strong location impression in this instance, that I did not pull through Wilkins' intention to take-off, or his actual take-off for Edmonton . . . even though I distinctly "saw" take-off of plane as well as its arrival.

7. My dream impression concerning contents of Wilkins' letter to me, beginning with apology, also confirmed. His two notes, contained in envelopes received February 14th, begin as follows:

"Sorry I have not been able to send you more messages by radio, but the short wave has been entirely off over most of the northern hemisphere during the whole of last month, and we have not been able to contact Iversen very often—and then it meant sending everything two or three times in order to get it through . . .

"There is an old saying, 'It never rains but what it pours.' With me it seems to be pouring all the time. *I feel quite ashamed* not to have wired you through the public office before this, and offered you some encouragement but every day I said: 'we are sure to get our radio signals back today.' They are still absent . . . we haven't had an even reasonably good connection with Iversen this year and have a pile of messages stacked up for him. . . ."

8. In my dream impression, I recorded Wilkins having written in a letter I was to receive: "always something coming up to interfere" . . .

In his letter to me, written February 10th, from Edmonton, and received by me yesterday, the 16th, he writes:

"I shall try to keep up my communication with you at 11:30 EST, but it is even more difficult here than at Aklavik, *for someone is sure to interrupt*" . . .

## Letter to Sir Hubert Wilkins Sent on to Sherman (As Recorded in Dream Impression)

216 Metcalfe St.,
Ottawa, Ontario,
January 22, 1938

Dear Sir Hubert:

First, I must introduce myself—Tam Deachman, the lad who is engaged to Florence Smith.

Florence told me of your interest in Mental Telepathy and Mental Pictures. This is a subject which interests me greatly, and I am looking forward to discussing some of these things with you if you come to Ottawa in the Spring.

I understand that it was your plan to do a certain amount of concentration at 10 P.M., Edmonton time. There was one day when you came through very strongly. There was no "mental picture," although there was a very strong mental contact. Unfortunately, I did not keep a note of it, but I know definitely it was either the 20th or 21st of December. I hope that if you were keeping any note at this time that you have a record of it, and confirm this. It came to both Florence and me very forcefully, starting shortly after midnight and lasting roughly for an hour. Although we were not together at the time, we compared notes the next day and found that we had both experienced your "haunting." I feel certain you would have records of it, because never have I felt anything so forcibly before. The times I mentioned are Ottawa time, and therefore would be around ten your time. Finally, it got a bit too much, and we both started fighting it. I wonder if you felt that also.

As far as I can learn from Florence, you are concentrating your efforts more on bringing about a control of mental pictures. Unfortunately, I do not get mental pictures, but I certainly do get hunches or premonitions, and I find that I am always very closely connected with Florence. For example, when she was sending you a Christmas card, I felt it very strongly. Again, you were in my mind very vividly. Consequently, I have been working more along the mental telepathy line. Do you know whether Professor Rhine's experiments at Duke University are dealing with mental pictures, or is he confining himself entirely to E. S. P.?

The Yogis evidently have this down to a science, but if they are dealing entirely with the mental, why do they put so much stress on physical training in order to acquire it? Surely there must be some short cut—some way to train the mind through mental, rather than physical exercises. Granted that they get results, but would it not be more direct to train the mind with the mind?

It seems that the whole secret lies in control of the objective mind, does it not? That is, if the objective mind could be controlled, so that it would not interfere with the subjective, once the desired impression is given the subjective, then the subjective would carry out its ultimate plan, that of externalizing the original thought. From what I understand from different theories, it seems to be established that mind can control matter, and if we take it a step farther, that matter is mind.

That is what I am searching for, the key to this control. Yoga seems to have it, as I said, but even they admit that it might take a life-time of physical as well as mental effort. If we are to believe the New Testament, it would seem that Christ accomplished it in twenty-five years. Evidently there must be a short-cut. I am so anxious to discuss these things at length with you, should you come to Ottawa.

Would you please "haunt" at about 9 o'clock your time on Wednesday, the 9th of February, providing this letter gets to you in time. That will be, I believe, 11 o'clock our time, and we will compare notes later. May you have the very best of good luck this year.

Yours very sincerely,<br>T. W. Deachman.

[It will be noted that this two-page typewritten letter, while not written by Wilkins, as recorded through the dream, does concern *two different* persons purporting to have psychic ability. I "saw" two enclosures in longhand which I interpreted to be from two different persons . . . in addition to the two-page typewritten letter . . . but the longhand letters were from Wilkins . . . the "conscious" confusion in pulling through this impression is easily understandable . . . and yet, the basic accuracy of this dream experience is also apparent.]

## Chronological Report of Wilkins' Check on Impressions Received from Night of November 15th, 1937, to, and Including, Night of February 3rd, 1938

As a preface to this report, let me record again my own mental and physical feelings as they seemed to mount from week to week during this period when I had no confirmation from Wilkins—no conscious knowledge as to whether a fair percentage of the impressions I was recording were right or wrong.

During all this time, I have been under terrific economic pressure—the necessity of getting certain writing work done and sold to meet current expenses . . . the necessity also of tackling and completing certain writing assignments which would not bring me an immediate financial return, and yet which would lay the groundwork for substantial returns later. In addition to these family responsibilities, I have kept religiously to the telepathic test task at hand, not missing a night of the appointed times, and making typed records of everything, which is not an inconsiderable job in itself—as this looseleaf book of more than two hundred pages will testify.

In January alone, I wrote sport stories totalling 45,000 words, and started and completed an entire musical comedy book for Broadway. I have hardly ever been to bed before one A.M. for weeks—and invariably have arisen around 8 A.M., with no rest periods during the day.

Last night, for instance, I retired at 3 A.M., and was up this morning at 8 A.M. I've spent the entire morning typing this material, am leaving in an hour or so for downtown, and will be out late tonight as well as tomorrow night . . . and so it goes.

My mind has been filled with other activities up to the very moment of my telepathic test periods—when I am compelled to put aside my conscious mind's concentration upon personal problems and ambitions, and open the door on this inner perceptive faculty.

In the year 1916, I had a serious abdominal operation—appendicitis with involvement of the intestines . . . This resulted in extensive adhesions. Occasionally, during the winter months especially, when my physical exercise is curtailed, I get "congestive kickbacks," resulting in gas pressure, kinks in the intestines, and other discomforts—such as severe sick headaches—all of this causing incessant pain in the stomach and intestinal tracts . . . much of which has to be borne for the time being.

I mention the above physical condition only that the scientific observer may get a clear personal picture of all the conditions that I, as a human entity, have had to combat in receiving these impressions.

Couple with this, also, my own sincere desire to meet every condition of testing as suggested by Dr. Gardner Murphy of Columbia University. I have felt more keenly than I have cared to admit my inability, due to circumstances beyond my control, to carry on more tests of a nature that Dr. Murphy and his staff might check. But Sir Hubert Wilkins has been unable to carry on short wave radio communication which would have permitted these tests to be held on regular schedule . . . only one, and even that in somewhat garbled fashion, has been completed to date.

Week after week going by with no confirming word from Wilkins, seemed to wind me up tighter and tighter on the conscious plane, which in turn made its increasing effect felt upon my body. Getting to bed as late as I did, I still could not sleep—an unusual condition for me—or, if I did drop off to sleep, it would be with such stomach and intestinal cramps that I would be compelled to get up and go into my study, remaining there the balance of the night, so that I would not disturb my wife—and thereby prevent her getting her rest, also.

My nerve condition, affecting my digestive apparatus, was also upsetting a chronic intestinal condition, made sluggish by adhesions . . . and the formation of gas caused me to be excessively bloated, with severe attendant pain. Dr. Seymour S. Wanderman and Dr. A. E. Strath-Gordon can testify to this condition—which reached its point of greatest severity a few days before receipt from Wilkins of my accumulated material, on February 14th.

So great was my relief, upon finding that my ratio of successful impressions had been upheld during this period . . . and even increased . . . despite these difficult conditions under which I have been laboring . . . that I could feel my whole body let go . . . a tenseness—caused by the conscious mind's wondering reaction to this whole process, so foreign to it in its so-called normal state of expression—slipped from me. I felt like a new person almost immediately . . . I drew the first completely deep breath in weeks . . . new inner assurance came to me . . . and my stomach and intestinal upset has largely cleared away . . . though I have had no more sleep, and have been even more intensely applying myself to creative work at hand—as my wife, and those associated with me can testify.

## Two Letters from Sir Hubert Wilkins Mailed from Edmonton *Re:* Tests

Aklavik,
Canadian Northwest Territories,
February 4, 1938.

Dear Sherman:

Sorry I have not been able to send you more messages by radio, but the short wave radio has been entirely off over most of the northern hemisphere the whole of last month; we have not been able to contact Iversen very often, and then it meant sending everything two or three times in order to get it through. I expect he has told you this.

I have had a message for you on file ever since the last mail, saying that I would try the cards in rotation on Wednesday, Thursday, and Friday nights, but since it has not gone off, I have not tried the cards at all yet.

I tried to get the annotations on your last batch in time to go out with the last mail, but as I received them only two hours before the mail left and there was a lot of other urgent matter to get off, I did not get it finished in time. We arrived back from Barrow a few hours before the mail plane left. It was unexpectedly delayed—it should have gone a week earlier.

You will notice that there are some "hits" in each of the tests you sent which is, I think, quite remarkable, as you could not have had any knowledge of some of the conditions. The fire at Barrow, for instance, although it was not much of a flare-up and the house was not burned down.

As before, you seem to get most, if not all the *strong* thoughts I express throughout the day—not necessarily at the appointed time. I am afraid I have not been very much on the job for the scheduled time comes just about our meal time here, and I am frequently out to dinner with the folks here and often forget about the schedules. However, some nights, I can remember to send some things and will annotate any messages you send in.

If we can get a good night on the radio in the immediate future, I will send out a message, in reference to the cards. I, too, think that unless we have a definite schedule on this, it would hardly be a fair test. I also think we might choose only five cards, one of each kind—concentrating on them in rotation—about one minute each one—then going over them in rotation once more. This also applies to numbers alluding to latitude and longitude. I will also notify you the scheduled time for figures, for I have not concentrated on figures as yet. Opposite the conditions referred to in your notes and where I have not yet made annotations, the impressions are not related to conditions here, or are not correct.

I have not remarked on conditions that were *nearly* correct, nor have I added any information about the conditions which might lead your "conscious" mind to color your impressions. In this way, it can not be said that you had any "leads." In fact, I think you have been working under the greatest handicap both from my lack of concentration and from the unusual conditions here.

You probably read of our one moonlight flight. The weather has not been good this year, but I hope it will be good for the February moon.

Word came today that the mail plane from Edmonton left yesterday, six days before schedule, but I rather think that another plane will leave on the scheduled date, and that we will get our expected mail by that.

Today, also, we received word of a plane forced down while on the way to Barrow for one of our men who was to go back to Moscow, and, as the plane is out of gas and we can make a search of the Alaskan mountains on our way to look for the local plane, we will start out tomorrow morning, weather permitting, for a flight. Therefore, I write this hurriedly tonight so as to leave it for the next mail.

Please give my regards to Mr. Emery, Mr. Whitmore, and other mutual friends, and with best regards to you,

From yours sincerely,
Hubert Wilkins

## Second Letter Received Same Mail, February 14th

Soviet Search Expedition,
Eastern Section
Aklavik
*Night of February 6th, 1938.*

Dear Sherman:

There is an old saying "It never rains but what it pours." With me it seems to be pouring all the time.

I feel quite ashamed not to have wired you through the public office before this and offered you some encouragement, but every day I said, "we are sure to get radio signals back today." They are still absent. We haven't had an even reasonably good connection with Iversen this year, and I have a pile of messages stacked up for him that are now (some of them) six weeks old.

There seem to be delays all along the line of this particular expedition. No skis, then no snow . . . then no moon, no cloud. And *then,* someone must go and put a stick up in the ice just a few minutes before we taxi down the river. We hit it with the propeller (without knowing it for sure) and fly for 9 1/2 hours. If we had had, or tried, to fly for 10 1/2 hours—we would probably be still *walking* home . . . for, *today,* the engine having run for less than an hour since we landed and only during "warming up" periods, went "hooey." We were just about to take-off. Lucky for us we were still on the ground. But it means a new engine, and I am not sure where or when we will get one.

Then, while we were checking up on the broken engine, the mail plane came in (six days early) with a batch of mail, yours among it. I still have your earlier notes here. I tried to get them off with the last mail which went a few hours after I received the parcel, but I was too late.

This time, although I had about two hours' sleep the night before last, and hardly any last night (have been carpentering, and building a shack over the broken-down engine all day), and it is now 4 A.M., I have to catch the mail plane for Edmonton at 8:30. I am determined to send you a few lines and put the notes and this scribble into an envelope, so as to be sure to get them posted as soon as I get to Edmonton. There will not be any time to do it when I get to Edmonton, I know, for I have to arrange about a new engine and what with phone calls, and so forth, I won't get any peace. But—enough of "bellyaching"!

You will notice in this batch of notes many "hits"—about as many as in the others. You seem to get *all* the very *strong* thoughts and sense the vivid conditions, even though I am generally unable to get down to concentration at the appointed time.

One reason I have not concentrated more is because, when I "concentrate," or try to, I think of a "million" things. I am as yet unable deliberately and immediately to *forget* all but one thing. I can concentrate, that is think solely of *one* essential urgent practical matter, but the other possibility has not come to me as yet—probably due to my spasmodic, and perhaps not very determined efforts.

There is always, or at least during the months since we met, some "upsetting" thing on my mind. I don't know what I would do if I had to attend to as many things as you attend to. However, the main point is that, in spite of the adverse conditions, there have been a *remarkable number of "coincidences impossible of reason"* or fore-knowledge, and that, while they may not serve the purpose we had in mind, they are I am sure, helping to develop your perceptibility.

I don't know if you got my message on Friday night referring to the ESP cards. It was the first time I really tried to do anything about them, although I have looked them over from time to time, and tried to guess which was which out of the five. I rarely got more than two, sometimes none, correctly.

The procedure on Friday, adopted by me, was to hold one card—one of the five different symbols—up before me for one minute, concentrate, closing my eyes, change to the next card, open eyes, concentrate, and so on. The order of the cards, picked after a general shuffle, was: star, cross, circle, square, and wave. I repeated these, over and over, for the half-hour of our agreement time ... 11:30 to 12 EST, and will be looking forward to the report on it, although I can't expect much without more deliberate practice.

Well now, I see it is time for another radio schedule. I'll have to rush—but will try to write some more in Edmonton.

                                Good night and best wishes and regards!
                                From—Hubert Wilkins

## Test 52

### February 17, 1938.     11:30–12:00 P.M.

*I tried an experiment tonight. Dr. and Mrs. A. E. Strath-Gordon came home with me from downtown, and they and my wife took seats behind me in my study, remaining throughout my sitting, which was done in the dark. I used a flashlight only to see notepad while writing down impressions.*

*The moment I had seated myself and focused my mind, I forgot their presence entirely, and tonight's results should not have been affected by their audience.*

*An exact transcription of tonight's impressions is as follows:*

| SHERMAN | WILKINS |
|---|---|
| You are thinking of me strongly tonight—made flight back Aklavik yesterday as I foresaw with new motor—flight made safely and without incident is impression comes to me . . . | Started back is correct. |
| Day spent in mounting new motor in place on plane—get feeling it is port motor—as I sit in plane—this operation seems to be arduous task . . . | Port motor. |
| Is motor transported across ice on large sled—especially constructed? See group of men pushing some large object across ice—and careful lifting operation—Eskimos or natives helping . . . | Not until next day. |
| New propeller to go with motor—seem to see it—something about pitch of propeller—seems as though pitch of both propellers must be same—I have no technical knowledge and do not know what this means . . . | Correct. Had to change pitch of one to suit the other. This done at Montreal, but I thought often about it. |
| Impression "68 north," as though you tried to impress me that you back in Aklavik—uncertain as to whether I knew this to be fact. | At Fort Simpson—thinking of you on schedule, but had no particular impression in my mind. |

## Letter to Sir Hubert Wilkins at Aklavik

Dear Sir Hubert:

Your long awaited packet of material, all carefully notated, has been received . . . and the painstaking scientific checking you have made is deeply appreciated by me.

I am glad you have given me no credit for other than "direct hits" . . . and feel, as you do, that an exhaustive analysis of all impressions, right and wrong, on completion of your expedition, should produce findings of definite value in the mental realm.

I am enclosing copy of letter written to Dr. Murphy concerning our one ESP test. The difficulty in rotation and a repetition of rotation is that I may tune in on the second card and follow through correctly, as I appeared to do—having missed the first . . . or I'm apt to break in on your series at any point . . . being actually a card or so late in the continuity. This is where the difference in time element enters in . . . and this kind of telepathic pick-up, it seems to me, almost calls for exact time synchronization.

However, this is only the first test . . . and I'm willing to attempt anything that seems feasible to you. Don't be concerned if you can't see the symbol or number in your mind's eye—just so you can *feel* it strongly . . . I think our emotional machinery provides the energy for sending . . . that's why I can pick up accurately so many things you have experienced. But the so-called future time element, which has "barged in" occasionally, is quite another thing—and has one guessing. We have evidence, however, that some law is at work here . . . or I could not correctly record what has not yet apparently happened . . . and is not in anyone's mind.

I am now busy making a summary of the material you have sent me—which is quite some job.

It seems necessary for ESP and similar experiments that I definitely know when you are sending. This enables me to eliminate my conscious mind's wonderment or uncertainty.

Whitmore, Emery, and I send regards and heartiest good wishes. It's high time for your luck to change—and hope I can commence picking up a host of good impressions! Thanks again for your swell cooperation under what I know to be great difficulties.

       Sincerely—

## Test 53

### February 21, 1938.  11:30–12:00 P.M.

*Tonight, we had Mr. and Mrs. Reginald Iversen as guests for dinner and read them Wilkins' report on the recorded impressions.*

*Iversen, New York Times radio operator, who has been frankly skeptical throughout these months, expressed great interest in many of the impressions, asking me: "Well, how in the world did you get an impression like that?"*

*It was Iversen's comment that many of the impressions were so personal and also so accurate, he didn't see how anyone could now deny that something unusual was happening.*

*Iversen gave his promise that, with radio conditions improving, he would arrange for further ESP tests.*

*I told Iversen to tell Wilkins, in this event, to select his cards and go through them but ONCE—it being my feeling that, at this long distance, when it is impossible to synchronize time exactly—a repeated going over of the symbols or numbers selected is almost certain to cause confusion in receiving. I may tune in at any point of the rotation . . . and continue with perfect reception for a moment . . . but, missing the first symbol or number would throw me out of the actual order. In the one test held, I apparently did just this thing—getting a run of three cards, after missing the first, exactly as Wilkins had selected them—star, cross, circle—and getting a wave, the last card chosen, in its exact relative position—last.*

*Since telepathy is the attunement of the receiver's subconscious mind with the subconscious of the sender . . . whether that sender be consciously sending or not . . . I believe the rotation of the symbols selected will be recorded in the subconscious of the sender, gone over just once . . . and my chances of getting correct order thus greatly improved. Repetition would only tend to muddle. At any rate, we intend to try such sending and receiving.*

*An exact transcription of impressions received tonight, is as follows. I again tried an experiment by letting Mr. and Mrs. Iversen and my wife sit in the study with me, while I recorded the impressions that came through. I will be interested to learn whether their presence in any way inhibited my "sensitivity." I seemed to feel Wilkins' thought strongly.*

| SHERMAN | WILKINS |
|---|---|
| *Installing of engine has been completed and testing of it carried on today—Very difficult job—feel that weather delayed your work one day . . .* | Cold south wind made for delay. |
| *Some one of crew seems to have hurt left* | Dyne had hands spotted with frost |

| SHERMAN | WILKINS |
|---|---|
| *leg during work on plane—someone else has skinned hand or finger . . .* | "burns" which blister or else the skin is pulled right off when the hand is pulled away after being frozen to any metal. |
| *See great clouds of smoke or vapor about plane and hear uneven coughs of motors . . .* | With a wood stove going inside the "tent" over the motor, there are always clouds of steam and smoke on cold days. |
| *Use made of part of damaged engine—see someone tinkering with it—removing some parts . . .* | Some parts of old engine fitted to the new. |
| *You have some wine with several friends who welcome you back Aklavik . . .* | They have had some liquor that I brought with me. I didn't have any. |
| *Your cold, which you mentioned in your letter to me, has given you a little trouble—in your head and throat . . .* | Cold really bad. |
| *You brought back to Aklavik several boxes of cigars, cartons of cigarettes . . .* | One box of one hundred cigars for Wilson. |
| *Seems as though Dyne wanted you bring him something special or some member of crew requested you to get some article—can't make out what it was . . .* | Kenyon wanted receiver radio. |
| *Someone has had toothache—sore condition mouth . . .* | I had tooth filled evening before I left Edmonton. Was still tender and jumped each time I trod heavily. |
| *Think you would like to get some word through to Iversen if you could reach him before Thursday—wonder if this thought in your mind tonight as you think of me? . . .* | Sent word to Iversen. |

In my report, mailed to Wilkins, about above "coincidence," I ventured a deduction which Wilkins later commented upon.

| SHERMAN | WILKINS |
|---|---|
| *My repeated reception of impressions at different sittings concerning the radio equipment Kenyon wanted seems to indicate the ability of a sensitive mind to follow a trend of thought in the mind of the "conscious" or "unconscious" sender as it pertains to one subject. Apparently this "radio" business for Kenyon had been one of the details Wilkins had been following through on and thus often in his mind.* | Correct.<br><br>Gave this intense thought. Kenyon wanted me to bring it in as expedition equipment, and so avoid "duty"—but I finally decided not to do this. |

## Confirmation Through Iversen of Impressions Received Night of February 21, 1938

On the night of February 22nd, the evening following the Iversens' presence in my home, they attended a lecture with us, given by Dr. A. E. Strath-Gordon at Steinway Hall, and I was shown, at that time, several radio dispatches received by Iversen from Wilkins, confirming impressions which had come to me in the sitting that the Iversens witnessed.

Tuesday and Wednesday are "off days" for Iversen in his radio work, so he was not to be at the *Times* until Thursday, ordinarily. He had come from his work, Monday evening, to my home for dinner. During the course of my sitting, this impression came to me:

> *Seems that you are not just satisfied with some mechanical operation plane yet—something still worrying you—think you would like to get some word through to Iversen if you could reach him before Thursday—wonder if this thought in your mind tonight as you think of me?*

Earlier that very night, Wilkins had gotten through to the operator on duty at the *Times,* two messages for Iversen, which he learned about the morning of February 22nd, and made a special trip in to the *Times* to pick up. Iversen stated to me that Wilkins has seldom put through a message to him during his "off-time" in all the time he has been North. I recall, when reading this impression to Iversen after I had finished my "sitting," his expressing some doubt about its being correct. He also expressed greater doubt when I read him this impression:

> *Installing of engine has been completed and testing of it carried on today—very difficult job—feel weather delayed work day . . .*

Iversen had gone on to say that the mounting of a new engine was a difficult job, and that he thought it would take Wilkins considerably longer to complete it.

Here is a copy of the radio message which had even then been received at the *Times* for Iversen from *Wilkins:*

Feb. 21, 1938
    ENGINE ON PLANE MOUNTED WON'T BE READY TILL THURSDAY . . .
    QTC I
                              LOP

I read to Iversen some of past impressions, now confirmed by Wilkins, and also read some of later impressions received, which Wilkins had not yet seen. One of these impressions came through on the night of February 14th, and was as follows:

> *You in communication Aklavik today—Kenyon seems to want more supplies or parts of equipment . . . You have to delegate New York, through Iversen, to secure some pieces equipment and rush through in relation to repair job on plane's motor—also in connection radio . . .*

Again, on night of February 17th, I recorded:

> *Radio attempt tonight which seems to have been more successful—report on operations Aklavik—need for several more pieces mechanical equipment . . .*

Iversen reports that "contact" was made night of February 17th, the most successful communication since Wilkins went North, despite magnetic predictions to the contrary. And the night of February 21st, when Iversen was present at sitting, I once more "tuned" in on Wilkins having done something about request of one of men for mechanical equipment:

> *Seems as though Dyne wanted you to bring him something special or some member of crew requested you get some article—can't make out what it was . . .*

This persistent thought, coming to me through three sittings, hooking up first with Kenyon in a very definite manner, the night of February 14th, was substantiated by a radio message received for Iversen while he was visiting me in my room, as follows:

IVN:

> UNLESS YOU ALREADY ORDERED HOLD EVERYTHING REF: KENYON'S RECEIVER HE WILL PROBABLY NOT GET UNTIL AFTER RETURN.
>
> WILKINS<br>10:10 P

Iversen told me, when turning over these messages, that Wilkins had ordered him, about the time I picked up first impressions regarding Kenyon, to secure radio receiver for Kenyon. He had been working on this since.

NOTE: It is significant to observe that this impression of Wilkins' desire to get word to Iversen was received by me while the Iversens were present at this sitting. Iversen was very dubious, saying Wilkins rarely tried to get in touch with him during his "off-days" . . . but Iversen learned the following morning that Wilkins had, this very night, radioed two messages to him through the relief operator.

## Test 54

### February 22, 1938.   11:45–12:20 A.M.

*I arrived home a bit late from downtown where I had been in attendance, with the Iversens and my wife, at a lecture given by Dr. A. E. Strath-Gordon.*

*I could feel Wilkins' thought in my direction, and was eager to sit down and "go to work."*

*The following is an exact transcription of my impressions:*

| SHERMAN | WILKINS |
|---|---|
| *Something seems to be cleaned out—as though a fine piece of wire is pushed through a slender piece of tubing—has something become clogged? I see an oily, messy job of some sort . . .* | Much of this to be done in fixing new engine. |
| *You are thinking of me more on schedule lately—if not whole appointed time, part of it . . .* | Correct. |
| *I seem to see you eat a steak tonight, with relish . . .* | NOT ME! No fresh meat here since January 1st. The wild caribou did not come this year. |
| *Morgan's name flashed to mind—radio* | |

the first series received . . . forget about the order . . . and start anew with recordings at the time when Wilkins was to repeat.

I have therefore listed three different sets of impressions. A study of these listings, comparing them with Wilkins' selections when the actual rotation has been radioed to me through the *New York Times,* should prove of great interest.

It will be observed that there has been a corresponding repetition of certain symbols in the same order, in a number of instances.

Only a check with Wilkins' selected list will determine whether this repetition was the result of a correctly received impression which persisted throughout . . . or whether the "suggestibility" of my own mind caused this repetition, once certain symbols were recorded the first time.

There is also the possibility of my having made a correct pick-up on my second or third try.

When it is realized, that even in short wave broadcasting, conditions have been so bad that it has been necessary to repeat messages two and three times, until they could be completely picked up . . . no less an opportunity should be given to the human mind in this pioneering work.

It is my belief that there would be less chance of confusion if the sender *did not send his selected symbols more than once.* I would much rather Sir Hubert Wilkins had concentrated three minutes each on each card selected, consuming 30 minutes in such a procedure, rather than having repeated a sending of the 10 cards twice. I think the continuity would have been more definitely recorded in his own subconscious and thus more clearly sent to me. Only further experimentation will tell.

## ESP Impressions as Recorded by Sherman in Attempted Synchronized Time Pick-Up from Sir Hubert Wilkins as He Repeated Tests

| 11:30 to 11:40 | 11:40 to 11:50 | 11:50 to 12:00 |
| --- | --- | --- |
| 1. Circle | 1. Wave | 1. Circle |
| 2. Square | 2. Star | 2. Square |
| 3. Cross | 3. Cross | 3. Cross |
| 4. Square | 4. Cross | 4. Star |
| 5. Wave | 5. Star | 5. Wave |
| 6. Star | 6. Circle | 6. Wave |
| 7. Cross | 7. Wave | 7. Cross |
| 8. Square | 8. Square | 8. Circle |
| 9. Wave | 9. Cross | 9. Star |
| 10. Circle | 10. Circle | 10. Circle |

## Wilkins–Sherman Special "ESP" Telepathic Test

February 25, 1938    11:30–12:00 P.M.

Through the cooperation of Reginald Iversen, *New York Times* Radio Operator, another appointment was made tonight with Wilkins for an ESP test.

This time, Wilkins agreed to select ten ESP cards, returning each card to deck after selection, for shuffling and re-drawing—and concentrating on each card for *three* minutes—*but not repeating* such concentration. In this manner, the half-hour period was exactly consumed.

I am of the opinion that this method, once we both get mentally accustomed to it, is going to be fruitful of interesting results.

To enable me to put my entire concentrative attention upon receiving of the symbol impressions, I had my wife help me, by sitting in my study and acting as time-keeper. At the end of every three-minute period, she would tap on the desk, and I would attempt to clear my mind for the next ESP card transmission. She sat at the other end of the room, and I sat with my back to hers at another desk, facing the wall.

There was a dim light on in her end of the room and I kept my eyes closed, with the palms of hands against them, looking within. There were moments when the sense of gazing into a great black void, alive with electrical energy, came to me. Spark-like myriad white or luminous fragments, ran crazily about in the inner darkness, and I had the feeling that they were trying to form into a picture of the symbols being concentrated upon by Wilkins.

Twice a symbol formed and could be seen distinctly in consciousness—once on number 7—when I saw a star . . . and again on number 8, when a cross appeared. Whether receiving of these impressions was synchronized with Wilkins' sending or not remains to be determined . . . but there is no doubt whatsoever of my actually seeing, in my mind's eye, these mentally formed images.

And here is a most important point—I am convinced, from this experience, if the receiver could induce such a mental state that he could look into a black void . . . and watch these images form, with a *natural conscious curiosity* as to what symbol would appear—he would be able to eliminate completely the intrusion of his conscious mind's speculative or imaginative faculty.

I found myself absorbingly intrigued in observing the tremendous movement of these spark-like particles which finally converged upon each other, formed a luminous irregular ball, and then startlingly took the shape of a star and a cross, on these respective occasions, without any conscious suggestion from me.

The sensation was the same as one would have received from watching this phenomenon projected on a motion picture screen. I felt entirely independent of what was taking place . . . the strain of trying to "pull in" an impression of what Wilkins was trying to send me, was gone.

I am recording my reactions as faithfully as is humanly possible in advance of my knowing what Wilkins' card selections were.

Whether I made an exact pick-up on these two occasions when I seemed to be able to retain this "black void state," is not so important at the moment as the technique that I believe I've hit upon for this kind of telepathic pick-up of *exact* impressions. It will require more practice for me to induce, at will, this state of consciousness—as it comes and goes—dependent upon will control . . . not a conscious, forceful willing . . . but a kind of directional surrendering to this state of inner darkness . . . I get a "sucked in" feeling, when this phenomenon occurs—an unconsciousness of my physical body—and I seem to exist in this blackness, surrounded by vibrant, luminous particles, trying to take shape. By looking at this dark field and holding the role of the observer, *willing* quietly but steadily that the image is going to form of which Wilkins is thinking—I am confident, in time, that a pick-up can be made, not once—but with fair regularity.

This point I should emphasize now—an experimenter or "receiver of telepathic impressions" must school himself to be impervious to mistakes. You cannot try to guard yourself against getting wrong impressions, nor can you bring your supposed reason into play. You must record what comes to you without any attempted discrimination.

The moment you become "self-conscious" regarding the receiving of impressions, "you are gone." The ego *must* not, *cannot* enter in.

I, myself, find it more difficult successfully to receive impressions when under observation. Something in me rebels against any suggestion of "showing off." I have been trying to overcome this "inhibiting reaction" by inviting certain friends to sit in the study with me during my appointed time with Wilkins. In the case of the Iversens, their presence apparently had no effect upon me. I brought through accurate and verifiable impressions. If I can continue in this development, it may subsequently be possible for me to demonstrate this telepathic faculty to observers who are entire strangers to me, when required.

It will be noted . . . in these ESP tests tonight, I have really made three experiments out of the one sitting. The three are as follows:

1. Attempted pick-up of the 10 ESP card symbols, one by one, as selected by Wilkins, at the exact time of his concentration upon each one.

2. An attempted pick-up of the entire 10 cards, as recorded in Wilkins' subconscious, following his completion of the test.

3. An attempted clairvoyant perception of the proper order of the cards, through a shuffling of my own ESP deck, and selecting ten cards in order, at random and on impulse, face down, just as Wilkins himself chose them.

Should any one of these three ways of receiving impressions be found to have been fairly accurate as compared to Wilkins' actual selections, it will throw considerable light on thought processes.

Next time I would like to add a test. I should like to see if "precognition" is possible—by sitting half an hour before Wilkins is to make his selection—and determine whether I can "foresee" the order in which he is going to draw the cards.

The observer will note that there are again "matching" relationships in the three different sets of impressions. Only a comparison with Wilkins' selected 10 ESP cards will tell how much can be charged to suggestion from my own mind . . . and genuine telepathic pick-up.

## ESP Impressions as Recorded by Sherman

In this test, Sir Hubert Wilkins shuffled the cards, selected one each, concentrated on this symbol for a period of three minutes, and returned to deck, shuffled and drew again, repeating this operation until 10 cards were drawn. He did not repeat the process.

This procedure did not hurry me in my attempt to record impressions being sent in a certain order before that order should be repeated, thus serving to confuse the receiver and muddle the continuity of impressions in the subconscious of the sender, due to uneven intensity of concentration on each card.

The three tests are as follows:

| At Time of Sending | After Period of Sending | Personal Selection ESP Cards from Deck |
|---|---|---|
| 1. Square | 1. Cross | 1. Cross |
| 2. Star | 2. Cross | 2. Cross |
| 3. Cross | 3. Star | 3. Wave |
| 4. Star (Cross?) | 4. Wave | 4. Star |
| 5. Circle | 5. Square | 5. Square |
| 6. Wave | 6. Cross | 6. Star |
| 7. Star | 7. Circle | 7. Cross |
| 8. Cross | 8. Square | 8. Square |
| 9. Circle | 9. Wave | 9. Wave |
| 10. Square | 10. Square | 10. Wave |

The only discernible difficulty in a three-minute concentration upon one symbol is the ability of the receiver to hold the mind blank, immune from intrusion of the conscious mind, thus spoiling the *picture* of the impression received . . . or the "feeling" that comes as a sudden, inexplainable "hunch" as to the symbol being transmitted. Also, if the apparent impression is received quickly, the mind must be kept from wondering as to what the next impression is to be until the time for selection arrives!

[After I had completed "session" and had left desk, turning on lights and walking out into other room, I suddenly got impulse to attempt another form of test . . . Taking my pack of ESP cards, shuffling, holding face down, and directing my inner mind to enable me to select the cards Wilkins chose, in the order he chose them. I have already recorded the results of this experiment. If I could have "taken on" the same mental processes used by Wilkins at the time, I might have been able to reproduce his own selections. Such an attempt as this is frankly experimental and pioneering, but many new phases and faculties of mind will be revealed in due course of time as a result of such investigation.]

## Report from Wilkins on ESP Cards Selected for Transmitting

February 26, 1938.

Tonight, around nine o'clock, came a phone call from Reginald Iversen, asking me to get pencil and paper—saying he had received a radio message from Wilkins, giving him order of ESP selections for our tests on Thursday and Friday evenings.

The order of these selections is as follows:

Thursday Night (Feb. 24)
1. Wave
2. Cross
3. Square
4. Cross
5. Wave
6. Circle
7. Square
8. Circle
9. Star
10. Cross

Friday Night (Feb. 25)
1. Cross
2. Wave
3. Cross
4. Square
5. Cross
6. Wave
7. Circle
8. Square
9. Circle
10. Star

Analyzing this actual order in which Wilkins attempted mentally to transmit the ESP symbols to me is a job for a mathematician.

On a rough check by me, I apparently scored three hits out of ten in both tests. I recorded my impressions three different times, under different circumstances, and my highest score in any one of the three was "three" rights.

In the first test, I got but one "hit" on the first sending 11:30 to 11:40, three hits on the second repeat sending, 11:40 to 11:50, and three hits on the next repeat sending, 11:50 to 12:00 midnight.

It is interesting to note, however, that I scored SIX hits out of TEN in a summary of the three attempts to receive Wilkins' transmission in exact order.

I mean by this—I placed six of the ten symbols in their proper order, correcting previous wrong impressions—or over-ruling them as I received correct impression of symbol during second and third repeat sending.

Wilkins' number 5 selection was particularly significant, since it was the only correct impression I received—that of a WAVE—on the first attempted receiving; I missed it on the second repeat, but got the WAVE again on the third sending.

I do not pretend to know what the percentages are—with Wilkins drawing always from 25 ESP cards, containing five each of the symbols. Chance, I understand, is supposed to be 5 out of 25, or less, in ordinary ESP tests. On this basis, without considering the different circumstances attendant upon these experiments, my reception was at the rate of 7 out of 25, or slightly better, which places it above the chance ratio.

I am by no means satisfied with these results and hope greatly to better them in the near future . . . but Wilkins and I have needed this kind of test to perfect a technique for the sending and receiving of just these kind of specific impressions.

Friday night's results gave me the same score—but the net result is better. I "hit" three times at the "time of sending," picking up Wilkins' transmission of a "cross" in third position, a "wave" in sixth position, and a "circle" in ninth position.

It is interesting to note my report made after completing this test—that I seemed to see a "star" and a "cross" vividly in mind in seventh and eighth position, "at time of sending." I *missed on both these impressions, but scored immediately afterward, in ninth position,* with a "circle." Whether the fact that I felt mentally keen up to this moment was any contributing factor, is a question. I'm still sure that this state of mind has a close relation to accurate receiving.

Recording my impressions of Wilkins' order of selection *after the period of sending,* gave me again a score of three hits but of symbols in relatively different position than the ones previously picked up. This again gave me *six* hits out of *ten,* compiling the results of the two recordings. I have no way of knowing the chances for or against my "hitting" the exact location of three symbols in ten. And I do not know whether Dr. Murphy would credit me with anything save impressions received "the first sitting," both times.

I would like to call attention to my recording, in parenthesis, a "cross" after my impression of a "star" in fourth position. As I had seen the star come into consciousness, it had been transplanted by a "cross," which impressed itself so strongly that I was compelled to put it down as a "second impression" for that position.

It is no doubt significant to observe that Wilkins' fifth selection was a "cross"... and he may have selected it while I was still concentrating on the "fourth place card." Then, having received the impression of a "cross," I had gone on to color my impression for fifth place by recording a "circle"... however, I tuned in on Wilkins' sixth place selection—a "wave" recording it correctly.

This might indicate to an impartial observer that I was close to the trend of Wilkins' thought from the third position, where I correctly named a "cross," to and through the sixth position, where I again scored with a "wave." I then hit again in ninth position with a "circle."

In the period after sending, when the symbols were in Wilkins' subconscious and he had stopped conscious sending, I got a "run of two," picking up his seventh and eighth position cards, *the very ones I thought I might have gotten* with my reported *star* and *cross* impressions received "at times of sending."

It is a coincidence that I should have mentioned the cards in seventh and eight position as seeming to be impressed more strongly in mind... not having received them correctly "at time of sending"... but, when it came to a repeat attempt to get these symbols correctly, despite the fact that I vaguely recalled in consciousness my experience of some minutes before in getting a "star" and "cross" impression, I should now be impelled to record "circle" and "square" for these same respective positions... getting the symbols *right* this time!

In my further experiment with the ESP cards, as explained in my report on the test, I also hit the eighth position symbol correctly, recording a "square." So seventh and eighth positions figured definitely in the results obtained.

Why should positions seven and eight have stood out so in consciousness—both of which I scored correctly on my second attempted pick-up, with a correct score on the ninth position on the first pick-up?

I am not trying to make out any case for myself. I am just raising these points and questions as an aid toward a more comprehensive analysis. And I shall await Dr. Murphy's report on these tests with great interest, also any suggestions he may have for future tests. Iversen, with the glee of a true skeptic, says: "Not so good, Harold, not so good!" and perhaps he's right—but I promise to do better before we're through!

## Test 56

### —With New ESP Test Added—
### February 28, 1938.   11:30–12:00 P.M.

Tonight, as a preparation for my ESP test with Sir Hubert Wilkins, I had my wife take our ESP cards and "put me through the paces."

She turned the cards over, sitting in the other end of the room. I sat with my back toward her, and recorded impressions as they came. A click of each card as she turned it down following her concentration upon it, enabled me to synchronize. I did not have her concentrate for the full minutes on each card as I had instructed Wilkins to do.

My high score was nine out of twenty-five twice. I hit two sevens and four sixes, dropping down to three when I seemed a bit fatigued.

I rested a few minutes, and then I had my wife concentrate on five cards, a minute each. My score was four out of five the first time. On three successive tests I was no worse than two out of five, employing this method.

I noticed an interesting fact, when we were checking results later, and my wife was reading off cards in the order chosen by her. As she was about to read symbols of which I had received a correct impression, a vibrant impulsive something inside me responded. I waited to hear her speak the name of the symbol that I knew she was going to call. It was almost as though a subconscious part of me, having once received an impression correctly, hung to it and tried to get it through to my conscious mind again, at the first opportunity. Much could be written on this point alone. If a "sensitive" could only hold this feeling . . . as a factor in the recognition of true impressions, he would seldom go wrong.

I decided to be in my seat at my desk in advance of Wilkins' testing time tonight . . . and to place my mind in tune, if possible. I sat from 11:15 P.M., and made the following recording:

### Wilkins Report of ESP Test Carried Out Night of February 28th

*(These impressions were received with Wilkins having shuffled deck, drawing, and concentrating upon 25 cards of ESP deck, giving one minute to each, starting at 11:30 P.M. EST.)*

| WILKINS | SHERMAN |
|---|---|
| 1. Cross | 1. Circle |
| 2. Wave | 2. Cross |
| 3. Cross | 3. Square |
| 4. Cross | 4. Square |
| 5. Wave | 5. Cross |
| 6. Star | 6. Wave |
| 7. Cross | 7. Wave |
| 8. Star | 8. Circle |
| 9. Circle | 9. Star |
| 10. Star | 10. Circle |
| 11. *Star* | 11. *Star* |
| 12. *Circle* | 12. *Circle* |
| 13. *Wave* | 13. *Wave* |
| 14. *Square* | 14. *Square* |
| 15. Circle | 15. Cross |

|  WILKINS      | SHERMAN     |
|---------------|-------------|
| 16. Square    | 16. Cross   |
| 17. Wave      | 17. Star    |
| 18. Circle    | 18. Square  |
| 19. Square    | 19. Circle  |
| 20. Star      | 20. Square  |
| 21. Square    | 21. Star    |
| 22. *Cross*   | 22. *Cross* |
| 23. *Wave*    | 23. *Wave*  |
| 24. Square    | 24. Star    |
| 25. *Circle*  | 25. *Circle*|

## Observations of ESP Card Test

It will be noted that I scored seven *direct hits* out of twenty-five, which is two above the chance ratio established by Dr. J. B. Rhine, originator of the ESP card tests, for extrasensory perceptive experiments.

The great difficulty of exactly synchronizing *time* over a distance of twenty-five hundred miles must be remembered and taken into account with Wilkins, as the sender, transmitting impressions of symbols at the rate of one a minute for twenty-five consecutive minutes.

It is to be observed that in addition to the number of "direct hits" scored, I was "one position impression late," in recording *eight* more correctly! I raise the question as to whether or not I had slipped out of synchronization with Wilkins' mind, on the time of pickup . . . and then made instantaneous contact during the "run of four straight impressions" and again at the finish when I scored "three out of the last four symbols" transmitted.

I have often noticed, even in home experimentation, with my wife or others sending impressions to me from one room to another, that I would frequently be one position late in recording a correct impression.

It is quite possible that the sender does not "release" all impressions at the same instant . . . and only "lets loose" or gives the impression over to the subconscious for transmission when putting his conscious mind on the *next* card chosen. If this is true, then it explains why the mind of the receiver is late in the pick-up.

I am positive that telepathic sending and receiving is done only by the subconscious mind—under the directional control of the conscious. But the conscious mind, in and of itself, cannot and does not project thoughts.

A study of this ESP card test will reveal the following:

> That I scored seven "direct hits," with a "run of four straight" from the 11th to the fourteenth position, inclusive, then a "run of two

straight," the twenty-second and twenty-third, and one, the twenty-fifth, or three out of four of the last symbols transmitted.

It will be interesting also to observe that I missed out on recording whole groups of symbols correctly by being "one position late" in pick-up, as evidenced by the following:

| SHERMAN | WILKINS |
|---|---|
| 4. Cross | 5. Cross |
| 5. Wave | 6. Wave |
| 8. Star | 9. Star |
| 9. Circle | 10. Circle |
| 10. Star | 11. Star |

*(Then I "hit" correctly for four straight)*

| 11. Star | 11. Star |
|---|---|
| 12. Circle | 12. Circle |
| 13. Wave | 13. Wave |
| 14. Square | 14. Square |

*(I missed out on next four when I began a "one position late pick-up again")*

| 18. Circle | 19. Circle |
|---|---|
| 19. Square | 20. Square |
| 20. Star | 21. Star |

*(At this point, I scored "direct hits" once more)*

| 22. Cross | 22. Cross |
|---|---|
| 23. Wave | 23. Wave |

*(And, finally, as half hour of sending closed, I was exactly synchronized with Wilkins' mind as evidenced by:)*

| 25. Circle | 25. Circle |
|---|---|

Had this synchronization remained throughout, the score would have been fifteen out of twenty-five, instead of seven out of twenty-five.

While this is no attempt to analyze or evaluate, such a task lying within the province of Dr. Gardner Murphy, the above observations seem marked and worthy of consideration.

## Confirmation by Wilkins of Impressions Received

*That night, February 28th, following the ESP Test, I continued sitting for impressions concerning Wilkins' search flight activities, and brought through the following:*

| SHERMAN | WILKINS |
|---|---|
| *Thought of glass comes to me—windshield—some device for keeping glass free of frost has been worked out—and installed...* | Frost screen on windshields installed two days ago. |
| *Weather has been no good for flying but you hoping better luck tomorrow—Wednesday, feeling definite concerning this day for action—Seem to see flight taking place this day over mountains—you can call it premonitory—but seem to be in plane experiencing bumpy air currents—looking down into deep pockets or valleys filled with snow and ice—dangerous flight...* | Got off on mountain flight at 9 A.M. Wednesday.<br><br>Correct. |
| *Get impression flights over mountains intended, despite certain risk, as warm-up for much more extensive Arctic flights to follow—taking full advantage of daylight and moonlight as it develops later in March.* | Correct. |

## Test 57

March 1, 1938.    11:30–12:00 P.M.

I was home all evening tonight, a rare occasion in itself these days, so approached the sitting in more leisurely and relaxed mood. Whether results attained give indication that this mental attitude contributes to greater accuracy in telepathic pick-up remains to be checked.

One thing I know, as I contrast the ability to make myself receptive without a "specific message in mind," there is no strain on the physical body. I feel confident that I can be influenced by Wilkins' thought, consciously or unconsciously expressed—and to "pull in" verifiable impressions from the fleeting series of mental pictures and "waves of feeling" which come to me—so long as I don't try to "force" such reception.

In ESP work, conditions of reception are different. The conscious mind tries its best to be of aid. It over-impresses the subconscious with its responsibility in "pulling through" an exact impression of a certain symbol. This over-anxiety brings about a reaction on the nervous system, which seems to center in the solar plexus.

If the sender, with every good intention, is not sending with sufficient "emotionalized force"—for want of a better way to describe it—then it is like trying to *pick up a weak broadcasting station with one's receiving set. Either you have to secure a more powerful set—or find some way of increasing the receiving power of the set you have—or the power of the sending station must be increased, to permit good reception.*

*Sometimes I find my body and mind becoming too tense, almost a subconscious realization that impressions from outside are too faint to grasp . . . my own system seems to try to "step up" its sensitized power . . . This either results in "unintended forcing," which colors impressions . . . or if I am able to "make contact," I cannot hold it long because of an indescribable nervous exhaustion.*

*Any individual who is doubtful of body changes or reactions during telepathic attempts at receiving thoughts—particularly when the mind is trained down to the point of picking up specific test impressions, should subject himself to a series of experiments.*

*I believe, as in all things, that continued practice brings greater mastery over all elements. A runner is more easily exhausted and is shorter of wind in the first weeks of training.*

*A human, learning to use the power of telepathy, long dormant in the mind of average man, is somewhat in the position of a person calling upon unused muscles, and inciting or compelling them to respond.*

*The mind itself seems to know no fatigue. It is only the instrument of the mind—the human body—that has difficulty in making the adjustment required.*

*Tonight, with no ESP test, my mind "went out" with a surge of human interest, to discover the present plans and experiences of Sir Hubert Wilkins, and the following is an exact transcription of my impressions:*

| SHERMAN | WILKINS |
|---|---|
| *Real action scheduled for tomorrow—400 to 600 mile flight in offing over mountains . . .* | Correct. 1300 miles. |
| *Investigating rumor reported by natives or Eskimos that an airplane motor was heard the day Levanevsky and companions came down—this motor over mountain regions—I feel tonight that this basis for your search this locality—Have wondered consciously why you would make flight this region—this seems to come to me as answer . . .* | Correct. |
| *Several flights, planned, shortly following one another—to cover mountain territory . . .* | Correct. |

| SHERMAN | WILKINS |
|---|---|
| *If Russian plane wreckage sighted, you would mark location on map—and make trek by land to reach scene...* | Correct. |
| *Believe cloudy weather over mountain passes holding up flight—You can't go too far in such weather—if you took off, you'd have to turn back...* | Correct. Wind drifting snow. |
| *There seems to be a certain mountain peak or elevation you intend to circle—in vicinity of which you believe Russians might be, if anywhere in this locality...* | South of Barter Island. |
| *Liquor—in that connection—seem to sense commercial interest—like some firm wanting endorsement—you considering if they will offer you enough money...* | Correct. Hiram Walker. Didn't offer enough money. |
| *I feel necessity for exercise of greater caution your part these next several weeks—have not experienced this feeling at any time before in many sittings I have had—deeply hope nothing may happen to force you down on any of your flights—Suggest you check oil and gas feed lines leading into engines as possible source of trouble.* | *Oil pressure on new propeller gave trouble next day!* |

## Test 58

### March 3, 1938.     11:30–12:00 P.M.

The Strath-Gordons drove my wife and me home from the Open Forum meeting of the Psychic Research Society tonight, and came up for a visit; but I excused myself at the appointed time, and went into my study.

I felt much like a radio operator tonight who hears his "call letters" coming in. I could seem to sense Wilkins' thought in my direction.

An exact transcription of my impressions is as follows:

| SHERMAN | WILKINS |
|---|---|
| *I had recorded my impressions on Monday and Tuesday of your taking off for search flight on Wednesday—Several times I had strange sensation of feeling myself in air with you—passing quite low over great shaggy, white and shadowy peaks—below me, gleaming stretches —and large dark patches where the sun's rays cut off—at other times, shadow of plane passing over snow . . .* | Correct. |
| *I have flashing picture of someone of crew pointing out several herds of wild animals as you passed over mountain valleys . . .* | Saw caribou. |
| *And several large flocks of birds taking flight because of your plane's invasion of their territory . . .* | *This was the first day this year* we saw snowbirds which come north in spring—Ravens and owls stay all winter. |
| *You exchange messages with Russian government today as result of flight and flights to come, having officially reported negative results your trip over Alaskan mountains . . .* | Correct. Reported flights over mountains completed. |
| *Dyne not quite satisfied synchronization two motors—they worked all right, but he seems think they can work even better—seem to see him tinkering with motors today . . .* | Correct. Altered pitch of propeller this morning at Old Crow after one hour's flight. |
| *Feel it your own opinion that Levanevsky and his comrades are down somewhere on Arctic ice—rather than in mountains, though you covering every possible area . . .* | Correct. See Press. |
| *You have gone out to dinner again tonight—and telling of a few of the observations made on your Alaskan flight . . .* | We have dinner at Old Crow Police Headquarters, 140 miles from Aklavik. |
| *More relaxed feeling from you tonight.* | Correct. Satisfied with mountain search. |

## Confirmation through News Story Sunday *New York Times,* March 6, 1938, of Recent Flight Impressions *re:* Wilkins

For purposes of direct comparison with impressions recently received and now confirmed by this news report, I am quoting excerpts from Wilkins' own story opposite extracts from my impressions recorded at different sittings and so annotated.

The multiplicity of impressions confirmed relating to so many phases of his activities is of especial interest.

Wilkins' story, radioed to the *New York Times,* is dated March 5, 1938, from Aklavik, N.W.T. The news account is headed:

3000 MILES FLOWN IN ARCTIC SEARCH
WILKINS CLIMBS MOUNTAINS TO BOUNDARY
OF ALASKA IN HIS HUNT FOR LEVANEVSKY

FINDS NO TRACE OF PLANE

CONCLUDES, IN SPITE OF ESKIMO REPORT,
THAT SOVIET PILOT DID NOT REACH COAST

It will be noted at once, by anyone carefully checking my impressions, that there are *four* confirmations in the news heading:

1. My feeling that Wilkins would get in several more search flights over the Alaskan mountains before Sunday, completing this phase of hunt.

2. That he would find no trace of the plane.

3. That these flights had been made, investigating the report of an Eskimo that he had heard the Russian plane over the mountains the day it was lost.

4. That Wilkins had concluded Levanevsky had not reached mountains and was down on ice.

## Impressions Checked Against Wilkins' News Account, Sunday *New York Times,* March 6, 1938

| SHERMAN | WILKINS |
|---|---|
| *(February 28th)* | (Excerpts His News Story) |
| Weather has been no good for flying, but you hoping better luck tomorrow— | On *Wednesday* we flew four times along the full length of the ranges on |

## SHERMAN

Wednesday, feeling definite concerning this day for action—seem to see flight taking place this day over mountains— *You determined make up for loss of time by getting in full schedule of flights in March—plan extensive action—seem to see 8 to 10 flights planned, crew know you mean business . . .*

(Tuesday, March 1)
*Real action scheduled for tomorrow (Wednesday) 400 to 600 mile flight in offing over mountains*—investigating rumors reported by natives or Eskimos that an airplane was heard the day Levanevsky and companions came down—this motor over mountain regions—*I feel tonight that this basis for your search this locality. Have wondered consciously why you would make flight this region—this seems come to me as answer . . .*

*Several flights planned shortly following one another—to cover mountain territory . . .*

(March 3, 1938)
*You anxious complete search flights over mountains next few days and finish canvass this territory in order to give attention Arctic ice flights with lengthening of daylight and approaching moonlight conditions—You would like to get in several more flights by Sunday if you could—busy atmosphere around your little camp . . .*

## WILKINS

easterly and westerly courses, and Thursday, after reaching Long. 153 degrees W., we hawked back and forth north and south across the mountains on courses less than ten miles apart until we reached the Alaskan boundary. We then made one more long west and east flight along the south slopes of the foothills before turning into Old Crow, where we had arranged to pick up gas before returning to Aklavik.

[Flights over mountains started on Wednesday in accordance those impressions received some days earlier] I believe that the search made by Russian and Alaskan aviators last Fall and that we recently completed in Alaska has covered that area with practical completeness, and that we must now conclude that Pilot Levanevsky, in spite of the *report by an Eskimo who claims to have heard his plane, did not reach the Alaskan coast. Had he crashed in the mountains, I think we would have seen the wreck.*

(From March 6 News Story)
[This news account tells of the number of search flights made by Wilkins, *all before Sunday, completing* search flights over mountains, as I had impression Wilkins planned to do, nights of March 1st and 3rd, "in next few days." These flights to be followed by Arctic ice flights, and confirmed by Wilkins' news statements.]

# Wilkins–Sherman Telepathic Tests

| SHERMAN | WILKINS |
|---|---|
| | We returned to Aklavik yesterday (Friday) and will fill our tanks with gas ready to be off for a long flight northward over the ice at the first opportunity. |
| *Think you are running into streak of better weather and will be able take off again soon . . .* | For the last several days we have been favored with excellent weather. Visibility was perfect throughout Thursday, and haziness enveloped the land only late in evening as we followed the Porcupine river to Old Crow. |
| *(February 28th) Levanevsky and his flying comrades—I still seem to see plane down on ice—wrecked—crew long since dead . . . (March 3, 1938) Feel it your opinion that Levanevsky and his comrades are down somewhere on Arctic ice—rather than in mountains—though you covering every possible area.* | (From March 6 News Story) It would seem evident that Pilot Levanevsky did not reach the Alaskan coast. Had he landed on the Arctic tundra on the Yukon flats it would probably not have been a completely fatal wreck—we'd have heard from the crew. |

## Test 59

### March 7, 1938.    11:30–12:00 P.M.

*Nirmal Das, Hindu mystic, visited me in my home this evening, and we had an interesting talk on mental phenomena, comparing reactions. Later, when it came time for my attempted "contact" with Wilkins, I invited Nirmal Das into my study. He sat on the couch during the half hour, while I sat at my usual place, back to him, at my desk.*

*This again was a test of my ability to make my mind receptive and record my impressions, despite the presence of another in the room. Making one's self "un-self-conscious" in this work is often quite a feat.*

*Only a check on what seemed to "come through" tonight can determine whether my having someone else in the room inhibited, in any way, my receptivity.*

*An exact transcription of my impressions is as follows:*

| SHERMAN | WILKINS |
|---|---|
| *Shift in good weather you have been having has, I believe, caused postponement of flight contemplated for today—but you standing by for flight tomorrow over Arctic wastes . . .* | Good weather forecast, but think fog at Aklavik. |
| *Some mail seems to have arrived Aklavik today or en route—strong feeling this connection and your thoughts of me as consequence . . .* | Mail arrived 3 days ago, March 4th. |
| *This next flight to be your longest to date—around 1300 miles out and same distance back . . .* | Had hopes.<br>Bad weather turned us back. |
| *Was tail of plane slightly damaged in bumpy landing during Alaskan mountain flights? Seem to see some work having been done in rear of plane . . .* | Skis and tail skid slightly damaged when taxiing at Old Crow. We ran onto a gravel bar in the river. Ripped bottoms of skis, but no repairs necessary. |
| *Feel you intend to approach within 150 to 200 miles of pole and follow along on an arc or line at about this distance—in thought Russian fliers may have come down around this latitude . . . Next few flights most dangerous of any attempted from standpoint of being furthest away from your home bases . . .* | Correct.<br><br>Correct. |
| *Flash of red—haven't had this in long time . . .* | Canadian Airways flying *red* cans gas from Old Crow to Aklavik. |
| *Some endorsement some advertised product—seem to see liquor bottle—which has been requested of you—believe you acting upon offer—make some decision on this today.* | Made decision some days ago regarding liquor advertisement. |

## Test 60

March 8, 1938.     11:30–12:00 P.M.

*I spent this evening at home, visiting with an old friend, Harry Loeb. He has been interested in mental phenomena for years and we experimented with ESP cards, with no startling results.*

*I have been able consistently to score between twelve and fifteen out of twenty-five in the clairvoyant test for ESP cards when concentrating in private, but find that this scoring is cut down considerably when I attempt to do it before "company."*

*I attribute this inability to demonstrate the same extrasensory perceptive prowess in public to the fact that people expect more of me—and it is almost impossible to avoid a "conscious effort to live up to what they expect."*

*Any "conscious trying" at all is fatal, as investigators know. The instant one tries, the imaginative or guessing faculty is called into play, and results average around chance or under.*

*By the same token, when an individual has been averaging chance results, and suddenly assumes a mental attitude which jumps his score from ten to fifteen out of twenty-five cards, or more, it definitely indicates that some other mental force has entered in. With a little practice, it becomes easy to tell when this extrasensory perceptive faculty is functioning, and when it is not. The mental feeling is distinctly different.*

*Tonight's sitting appeared to be a blend of clairvoyant and telepathic phenomena—with certain premonitory flashes mixed in.*

*An exact transcription of tonight's impressions is as follows:*

| SHERMAN | WILKINS |
|---|---|
| Fleeting vision of your face—quite a strained, intent expression as though concentrated upon flight activity in plane—Barter Island comes to mind—seems as though flight started and down at some point or turned back—plane motionless—something did not go today as planned—snow or sleet-like weather some parts, seem to see it pelting plane . . . | This all happened three days later, on 11th March. |
| I see you beside plane looking up and around—there appears to be a ridge or slope beyond—of snow and ice—in ribbon-like outlines—There is activity about a camp or tent not far distant—some men moving in and about . . . | River between land covered with stunted trees—look like ribbons.<br><br>On 11th March. |

| SHERMAN | WILKINS |
|---|---|
| *I caught glimpse of a plane partially submerged in water and drift ice—one wing especially clear—left wing—greyish blue metallic color . . .* | On 11th March. |
| *Radio activity today, seemingly connected with plane—seem to see sparks dancing in darkness—a man with headphones on—very busy—Slight break in clouds above—but dark storm clouds low on horizon—intensive work just ahead—five big flights planned to encompass wide expanse Arctic ice territory—Strange feeling in pit of stomach—or solar plexus—like I've gone through close escape or acute experience—restless sensation also—trying to account for it.* | All on 11th March.<br><br>11th March.<br>Landing with full load—Taxiing badly. |

## Test 61

### March 10, 1938.   11:30–12:00 P.M.

*This is usually the night I attend the Open Forum meetings of the Psychic Research Society at 71 West 23rd Street, and listen to addresses by Dr. Henry Hardwicke and Dr. A. E. Strath-Gordon.*

*I went to the meeting as usual, and, while seated there, around 10 P.M., felt a strong "contact" with Wilkins . . . together with an urgent feeling to be home on time to keep my telepathic appointment with him.*

*I came up in the subway with my wife and Dr. Strath-Gordon, bidding him good-night at the door of our apartment house at 11:25, and hurrying upstairs. The doctor often parks his car outside our apartment house for the evening, and drives off to his home in New Jersey from this point.*

*Taking just time to take off my overcoat and hat, I closed my study door and sat down to my desk, taking out pencil and notepad. Impressions were even then starting to flow, and I commenced recording them—this time without turning out the light. I wrote almost continuously for the half-hour period, seeming to feel a contact throughout . . . and turned off the light only during the last five minutes of the time.*

*I shall be greatly interested when a check is made on the accuracy of impressions received this evening, because of the somewhat different conditions surrounding the manner of their reception.*

*An exact transcription of impressions received is as follows:*

| SHERMAN | WILKINS |
|---|---|
| You seem to be about to make one of the greatest search flights of its kind ever attempted in North—a flight which will require combined light of sun and moon to supply visibility—giving you opportunity of surveying, observationally, a wide and carefully charted expanse of Arctic sea. | Attempted long flight—turned back account weather. |
| Services of both pilots will be required on this flight, changing off. I get feeling you will probably be in air between 15 and 18 hours—covering between 2200 and 2600 miles . . . | On the 14th and 15th. In air 19 1/2 hours—3020 Miles. |
| "Diamond mine"—why I should think of this is mystery—almost as though you had discussed possibility existence of diamond deposit in Alaskan mountain area or north region—To my conscious mind this seems absurdity, but I record impression nevertheless . . . | That night was telling people at table, at about the time you were sitting, of visit to African diamond mines—discussing Cullinan and Jonker diamonds, and so forth. |
| I get glittering diamond-like impression with it—Is there a small scar on the left side of your face near line at edge of mouth? | Yes. |
| "Point Barrow" comes strongly to mind—some activity connection this base. | Point Barrow was sending radio direction signals that day. |

## Observations and Confirmations Concerning Recent Activities of Wilkins in Far North

March 12, 1938.

These past several weeks have brought to light a number of interesting revelations as to the "time dimension" in the receiving of impressions.

Now that returning sunlight is permitting Sir Hubert Wilkins to fly more often, and he is planning a number of flights—I have found it increasingly difficult to distinguish between mental pictures of planned flights and flights actually made.

A visualization of future flights exists in consciousness, as well as a mental picture record of flights already made. Usually, the only detectable difference between impressions of the past and future is the difference between their impact upon my emotional system.

I ordinarily regard a more strongly felt impression as pertaining to an event that *has happened,* since it is most apt to have aroused a more intense emotional reaction, registered in the subconscious of the sender, than has an event anticipated.

Yet, seeing a flight in my mind's eye, I have to decide in that hair-split second of so-called time, whether the flight has been made, is being made, or is to be made.

The mental sensitivity of an individual who has schooled himself through months of regular sittings for impressions, must be taken into account. To describe accurately the feelings and the nature of this sensitivity to one who has not experienced such reactions is most difficult.

I have had headaches at different times, some of which I have recorded, which I am sure have been caused by congestion of different nerve centers, brought about through telepathic work. Stomach and solar-plexus reactions have also occurred. In these instances, I am just as certain that I have permitted my body to grow too tense; I have been over-eager to receive accurate impressions; I have let myself be too much affected by assailments of doubt from conscious mind; I have had waves of being too much impressed by the increasing responsibility of meeting any tests that might be imposed, with the urge to produce more and more convincing results.

Added to all these factors is another very dangerous one where effective telepathic communication is concerned, and a point I wish to make clearly again.

## Danger of "Conscious Knowing" or "Expectation" of Certain Information As a Guide to More Specific Tests

Some weeks ago, as outlined in my report, I endeavored to get through a radio message to Wilkins containing a suggested specific test by Dr. Gardner Murphy.

Reginald Iversen, radio operator, told me that he would put the message through on a certain date . . . and for several sittings thereafter, my conscious mind colored my impressions by trying to tell me that Wilkins had received the message, and might, at each sitting, be telepathically trying to cooperate as suggested.

It is significant to note that, after I had gotten this "conscious coloring out of my system," *I went on to record accurate telepathic impressions as usual.*

Recently, within the past two weeks, I mentioned to Iversen that another experiment I should like to try would be the attuning of my mind to Wilkins during one of his actual flights—keeping a chronological report of my impressions.

"Fine," said Iversen. "Next time I learn that Wilkins has taken off, I'll let you know . . . and you can see what you can pick up."

Since Iversen made this statement, my *conscious* mind has anticipated a call . . . and has fought my subconscious self, during each sitting, trying to "tell" me that "Wilkins must not have taken off yet, else Iversen would have notified me."

This mental condition, while it has not produced as much coloring, has retarded, to a degree, my receiving of accurate impressions.

*I still have had no word from Iversen,* and am doing my utmost to "kill off" any conscious wonderment or anticipation of Wilkins' activities. I think very largely that the success I have obtained to date has been due to the fact that, during most of this testing period, I have kept my conscious mind free of any speculation concerning what might have been happening to Wilkins—and have only opened my consciousness for the reception of impressions relating to him at the appointed times. The fact that I have been so intensely busy with other mental matters—my own creative work—has been a great aid.

## Comments Concerning Wilkins Flight Made Last Thursday: My Impressions Recorded on Thursday Night, All of March 10th

For the first time since Wilkins has been north, I did not definitely record the flight he had made the day of my sitting. I have usually recorded an impression of his being in the air, if he has flown on the day of a sitting.

This night, as my prefaced remarks will show, I "strongly felt Wilkins thoughts concerning me," and could hardly wait to get home to start recording my impressions. As I sat down and grabbed up my pencil and notepad, I had a fleeting "conscious combative thought," still plaguing me relative to Iversen's not having notified me of a Wilkins' take-off; therefore, I could not be certain he was still in the air.

Whether this tempered what came through, and influenced me despite my usual attempt to control and eliminate my conscious mind from any consideration whatsoever, can only be determined by events in future time.

It will be noted that I recorded, at once:

> *You seem to be about to make one of greatest search flights of its kind ever attempted in the north . . .*

At the moment of recording this impression, I saw Wilkins in flight in my mind's eye—and almost found myself writing, "you are just completing one of greatest and so forth" . . . but impression of future time seemed a bit stronger, and I recorded as above.

My mind became so filled with thoughts of this flight that I did not record fact of flight then having been made. Wilkins, according to the news story in Friday morning's *New York Times* (March 11th), must have returned to Aklavik from this 2,080 mile flight, shortly before my "reception period," allowing for the difference in time.

I felt he was thinking of me, trying to impress me with certain facts, as I recorded. I am led to wonder whether, if he had not been trying consciously to impress me, would I have picked up his thoughts more easily (should this have been proved to be the case)?

Even so, I scored several direct hits, as will be noted:

| SHERMAN | WILKINS |
|---|---|
| *(March 10, 1938)* | (News Story, *Times,* March 11, 1938) |
| *Servicing plane for take-off impression . . .* | We landed at Aklavik shortly after 6 P.M. and S. A. Cheeseman and A. J. L. Dyne are now filling the tanks with gas so that Kenyon and I might take off again tomorrow, weather permitting. |
| *Work on plane impression—busy scene . . . Stove on ice—feel warmth from it, talking, intent on something—you look up at sky speculatively—and shake your head—Kenyon gazes at sky thoughtfully also—glances at Cheeseman, consultation some sort, indecision sensation.* | (While no confirmation, as yet I believe this "scene" impression, in light of above described activity, will be found to have been accurate. Kenyon discussing proposed Friday flight, "weather permitting.") |

It must be remembered that I am deliberately keeping my mind off Wilkins, and am not attempting to pick up any telepathic impressions during my daily activities.

The average person makes a few *conscious* pick-ups of thoughts with which he is constantly being bombarded—being protected by this *same conscious* mind which serves as such an "interference" when one tries to get telepathic impressions.

It ordinarily requires a developed *receptivity* of mind to be certain of "outside impressions," as distinguishable from impressions originating within one's own mind. This is the reason, in addition to the fact that I have been tremendously busy with my own creative work and other activities, that I have disciplined myself to give little or no thought to Wilkins except at the appointed telepathic times. It is humanly difficult enough to avoid "coloring," as it is.

## Corroborative Evidence of Prevision or Premonitory Impression Concerning Second Accident to Wilkins' Plane!

It will be seen, in checking back over my impressions, that I had several definite premonitory flashes concerning a "mechanical mishap" occurring to Wilkins' plane, and on the night of January 27th, recorded the following: (this being impression of "first" accident)

> *Have impression attempted take-off around February 11th—marred by incident which will occasion further delay—again I write this as though impelled, and tuning-in momentarily on a future condition . . .*

On the same night, apparently coupled with the above recorded impression, but separated by several other impressions, I put down:

> *Crank case of plane comes to me suddenly—did something go wrong with it due to cold?*

I then followed this recording with a very accurate impression of a "dead dog" on the ice, which Wilkins later confirmed having seen that day.

And on February 10th, following through on the premonitory phase of the above impressions, before I had received any word from Wilkins, I recorded:

> *Feeling crowds in—impression that came to me several weeks ago—as though unexpected development—mechanical mishap this time perhaps— would compel still another delay—I sincerely hope not, and this type of impression has no relation to faculty of telepathy, but I record it here because it came through . . .*

At the time of this recording the accident had already happened—on February 6th—and when I heard from Wilkins by mail on February 14th, he had annotated on the copy of my January 27th impressions, opposite my recorded *feeling* concerning trouble with *crank case* of plane:

> *Wilkins: "Wonder if this was 'forethought' or preview knowledge since it was not until February 6th that we developed serious trouble in crankcase—main bearing of one engine was ground to powder that day. Must have been some trouble there since January 15th."*

I mention again, in order to list all premonitory impression of unusual significance in relation to each other, that January 15th was the date I had *seen* Wilkins taking off on his next search flight—this impression received as early as the night of December 28th, when I had recorded:

> *January 15th comes to me as day you actually make take-off for north regions—though you hope to get off few days earlier in month—this again is premonitory impression—as though thoughts jump ahead of present moment—and this future moment, for a fleeting instant, becomes now—then fades again . . .*

Once more on the night of January 13th, I recorded:

> *Throughout the day, unbidden, has come the flashing impression that weather conditions again are causing postponement of flight intentions—am again impressed with the 15th of January as your probable take-off date—that conditions won't permit attempt till then, despite your present standing by attitude . . .*

Wilkins, writing of my impression concerning flight on January 15th, received night of December 28th, wrote:

> *"Quite extraordinary that we did make the flight, starting night of 14th (the 15th your time), and returning on the 15th of January."*

## And Now "Second Accident Premonition" Confirmed

Ability always to classify properly impressions received in relation to time—past, present, or future—is difficult.

Sometimes, as in the cases referred to above, I have been able to recognize the impressions received as definitely premonitory.

In this most recent instance, my recognition of a "future event" impression was not so clearcut. I indicated my uncertainty of the "time element" involved, however, by recording my impression in the form of a question, as follows:

(Night of March 7th)
*Was tail of plane slightly damaged in bumpy landing—during Alaskan mountain flights? Seem to see some work having been done in rear of plane.*

On the following night, March 8th, I apparently dipped in upon a premonitory series of impressions again, for I recorded:

*Seems as though flight started and down at some point or turned back——plane motionless—something did not go today as planned—Snow and sleet-like weather some parts—seem to see it pelting the plane . . .*

*Slight break in clouds above—but dark storm clouds low on horizon . . .*

*Fleeting vision of your face—quite strained, intent impression as though concentrated upon flight activity in plane . . .*

The above impressions all appear to have had a definite premonitory relation to an event which happened Friday, March 11th, and was published in a news dispatch in the *New York Times,* March 12th, wherein Wilkins tells of damage done to the *tail-skid of the plane in a forced landing,* upon the approach of a storm, shortly after their take-off!

It is important to note that, not once, in all the hundreds of recorded impressions, have I mentioned any damage befalling the "tail of the plane" until this time—just a few days before its *actual happening.*

That my impressions may be studied in direct relation to Wilkins' news story of the accident, I repeat them.

## Confirmation of Premonitory Impressions Relating to Second Plane Accident

| SHERMAN | WILKINS NEWS ACCOUNT |
|---|---|
| (Night of March 7th) Was tail of plane slightly damaged in bumpy landing—*during Alaskan mountain flights? Seem to see some work having been done in rear of plane* . . . | (Morning March 12th) WILKINS' PLANE DAMAGED Ship Hits Snow Ridge Taxiing, and Fuselage is Torn. AKLAVIK, N.W.T. March 11. (By Wireless) . . . |
| (Night of March 8th) Fleeting vision of your face—quite a strained, intent expression as though | An odd freak of the weather today led to the *second* accident our airplane has |

## SHERMAN

*concentrated upon flight activity in plane—Seems as though flight started and down at some point or turned back—plane motionless—something did not go today as planned.* Snow or sleet-like weather some parts—seems to be pelting plane—*Strange feeling in pit of stomach—or solar plexus—like I've* gone through close scrape or acute experience—*you concerned about something...*

*Slight break in clouds above—but dark storm clouds low on horizon—"Carry supplies"—these words come to me and feeling as though supplies of some kind transported...*

(Night of March 8th)
*I see you beside plane, looking up and around—there appears to be a ridge or slope beyond—of snow and ice—in ribbon-like outlines...*
*There is activity about a camp or tent not far distant—some men moving in and about...*

## WILKINS

sustained during our efforts to locate the missing Soviet airmen.

It was light and clear this morning until 6 o'clock, but shortly after we took off, a snow-laden squall, as black and sudden as a thundercloud, enveloped us, and fearing heavy snow and "icing up" on the machine, we were forced to land while we could.

Pilot Herbert Hollick-Kenyon made a good safe landing with our heavy load of 1,200 gallons of gas and equipment, but in *taxiing back to our starting point,* we struck a solid sharp ridge of snow, and the *tail-skid was torn from the fuselage.*

Engineers A. T. L. Dyne and S. A. Cheeseman, whipping back and forth between the machine and our main base in a home-made, propeller-driven sled, are quickly effecting repairs, and by working throughout the night expect to have the machine completely repaired by tomorrow.

[A checking of the Wilkins' news story with these recorded impressions will reveal their remarkable similarity throughout, even to the "carry supplies" impression, relating to work being done by Dyne and Cheeseman in taking supplies from camp to plane on sled in order to make repairs.]

I even received advance impressions of Wilkins' physical reaction to accident and perilous moment.

My pre-vision of damage to tail of plane was exactly described as taking place through "bumpy landing..."

The weather was accurately described, even to "dark storm clouds on horizon... snow and sleet."

I "saw" plane turned back on flight . . . something not going as it had been planned.

I failed only in not definitely cataloguing these impressions as premonitory . . . but their accuracy cannot be doubted . . . nor the genuineness of the pre-visions.

Such impressions as the above open up an entirely new field of speculation and study concerning mental phenomena.

What *is time?* This is an important question in relation to human consciousness and human concept.

Were the above events, pre-visioned by me, even then destined to happen? Had all of their causative forces been set into motion? Had the humans involved only to attune themselves to these causative forces at a point in time in order to materialize them—to bring them within our concept of time—making them a part of our present moment and then our past?

I will have more to say concerning my thoughts on these premonitory phenomena later.

## Radiogram from Sir Hubert Wilkins at Aklavik

March 13, 1938

On this date, Sir Hubert Wilkins, while in radio communication with Iversen, *New York Times,* sent me the following message:

3 RUPUL-A AKLAVIK NWT MAR. 13, 1938 SHERMAN NY
DO NOT THINK SO MUCH OF CARD TESTS
EACH NIGHT I REVIEW DAYS EVENTS REGARDS
WILKINS 710P

This message to me is significant. While I have been willing to try as many ESP card tests as could be arranged, Wilkins has no doubt discovered, as have I, that general reception of impressions, *emotionalized through human experiences of one kind and another,* carries a higher degree of accuracy and more specific results than the well-intentioned ESP card tests.

To the best of my knowledge, no one has ever attempted such sending and receiving, under such control conditions and over such a long distance as Wilkins and I have been conducting for the past five months.

For this reason, what we have found to be true and effective must be taken into consideration.

I feel that Wilkins has received more mail from me at Aklavik, and has had the opportunity of studying my report of ESP card tests. He has probably observed that, regardless of my score in the ESP tests, I have gone on—the very same night—to record accurate impressions of his activities for the day—proving that one's ability to pick up "emotionally charged" impressions is much greater than the more difficult and more inhibited test of "bringing through" impressions of symbols.

I am certain if the sending of symbols could be made a part of an individual's human experience, the receiver would show an immediate improvement in his score.

For instance, if the sender of ESP card impressions was subjected to sudden pain as he gazed at each card, so that his emotions were stirred in association with his concentration upon a symbol—this might accomplish a greater "sending intensity," approximating the same type of sending intensity which normally and unconsciously goes into a recording of each human experience.

Even then, there is a certain artificiality to such a procedure which is not present when a human is about the business of living.

It must be kept in mind that Wilkins is living intensely every day. Many days he is facing actual dangers—a multiplicity of problems. Consequently his consciousness is kept alive with a myriad of emotions and thoughts. This is why he affords such a good subject for these telepathic tests.

An average person, living in an average humdrum community, would record events in consciousness mostly on the same emotional level . . . and there would not be enough variation from day to day to give the receiver clearcut individual mental picture impressions or strong "feeling reactions."

This is why humans generally, when they *do* pick up a telepathic impression, find that it usually has to do with some tragic event in another person's life which has so raised the emotional level that a strong radiation of the experience has occurred . . . breaking through the ordinary conscious resistance to external thoughts maintained by the receiver.

*Iversen explains in note accompanying copy of Wilkins' radio message, why he did not notify me of Wilkins' flight take-off so that I might try new test.*

My conscious mind was finally put at ease by the following note from Iversen:

Dear Harold:

Been extremely busy on the new transmitter, and haven't had much time for anything else.

Had a good contact with Wilkins tonight (Sunday). His nights have all begun in the early morning, and he usually returned before our schedules, so couldn't advise you when he was in the air.

Sincerely,
Reg.

Only those who have attempted to receive telepathic impressions on a regularly scheduled basis, week in and week out, can know the feeling of physical and mental relief which came over me on receipt of Iversen's message.

I was determined that I would not phone him and make any inquiry. I have never asked him for information, and he has been very scientific in every contact I have had to make with him in connection with his work. Never has he given me any knowledge or explanation of Wilkins' work or activities, nor have I asked—as I am sure he will testify.

I have definitely not wanted any cues or leads of any kind, for I know how damaging they can be as my *conscious mind* attempts to impress my inner consciousness with what it regards as "known facts" as opposed to impressions I may be receiving, containing just the opposite information.

These experiments will show conclusively, on the occasions when the conscious mind has been "keyed up" by some expected line of action, how difficult it has been for me to rise above my conscious mind's interference, and thus prevent "coloring" from taking place.

This faculty of telepathic reception has no connection whatever with the conscious mind's wondering, guessing, reasoning, calculating, or anticipating activity. Unless such mental trends are "killed off," or "blanked," or suspended entirely, it is impossible to get accurate, verifiable impressions from without.

It can readily be seen how wrong concepts formed by one's conscious mind, and accepted by the individual as facts, can alter or ruin a person's life . . . just as they can obliterate a true impression received from the mind of another . . . or from the intelligence apparently existent in this time-space dimension about which we yet know so little.

## With Respect to ESP Tests

I must accept Wilkins' judgment on the ESP card tests, and set myself to the continued task of attempted reception of general impressions concerning specific happenings relating to Wilkins and his expedition.

It is quite possible that his explanation: "Each night I review the day's events" is accountable for my great number of accurate pick-ups. Certainly this kind of mental reviewing is proving a help.

## Test 62

### March 14, 1938.     11:30–12:00 P.M.

*I spent a quiet evening at home tonight, which was an innovation in itself. As a consequence I did not approach tonight's sitting like a fireman en route to a fire.*

Sitting in my study, at 11:30 P.M., I had a strange feeling as though Wilkins stepped into my room, exactly on schedule. I shall be interested to know if he was thinking of me exactly on time tonight.

I should also like to know some day—if it is possible to project one's entity into so-called space—at what point between, the minds of two communicating persons meet! This is not such a ludicrous speculation as it might first appear, judging from experiences I have had. This point of contact would vary under different conditions of sender and receiver . . . and sometimes would not occur at all . . . but, on occasion, I am convinced that more than an impulse is transmitted. When this happens, an entity momentarily seems to have direct contact with a distant locality—being able to come away with an exact impression of a place hitherto unseen, and without the aid of any other consciousness. This again is a different type of phenomenon, and must remain, for the time being, in the uncharted area of the human mind.

An exact transcription of impressions received tonight is as follows:

| SHERMAN | WILKINS |
|---|---|
| There has been plane activity today, but tomorrow into Wednesday, weather permitting, seems to me to mark one of your greatest attempted flights to date—making use of moonlight as well as daylight—feel some such plan strongly in mind . . . | Correct. |
| Work on replacing tail-skid appears to have taken a bit longer than expected . . . | 12 hours. |
| Believe you discovered crack of framework in tail or fuselage which also needed repair—somewhat more damage caused than originally thought . . . | Bulkhead was cracked. |
| I seem to see you manipulating a hand pump of some sort in flight—one of engines emitting black spouts of smoke—sharp detonation of motor—uneven—choked sound—as though some carburetor trouble—gas feed . . . | We use a hand pump to "strip" the fuselage tanks.<br><br>It was used on the long flight. |
| Ice on plane—thin coating which you watch closely . . . | There was at one time a very thin coat of ice. |
| See plane circling low over certain | Plane came down to within 1,000 feet |

| SHERMAN | WILKINS |
|---|---|
| area—icy waste with several open stretches of water . . . | of ice, and while at that altitude crossed several open leads. |
| Impression 86°—115—location | We were at 86° 115' W, and our turning point after careful checking was 86.50 N, 105 W . . . 105 W is only 45 miles from 115 at that latitude. |
| See another flight turned back because of weather conditions turning against you— and strenuous effort made to reach base before snow and sleet close you in . . . | We turned back because we were at the limit of my intended outward flight. But Kenyon wanted to turn back earlier, because we had information as to bad conditions at Aklavik. |
| March 18th flashes to mind again as a date of considerable significance concerning you . . . | Reached Edmonton 18th March. |
| Preparations going forward for outstanding two flights of expedition—one especially, designed to top them all, if weather and operation of plane permit . . . | Preparations made and strong in my mind—but the flights did not come off. |
| Seem to see spots of oil or something on windshield or glass window—did some connecting rod or something come loose while warming engine? Can't exactly account this impression. | A few spots of oil were on my windshield which annoyed me, and I tried several times to get them off, but could not reach them. |

## Test 63

### March 15, 1938.   11:30–12:00 P.M.

*Again, I was at home all evening. Some of the creative writing pressure has been off me for several days. There are other kinds of pressures, however, in pending contracts and the mere economics of living in these strangely distorted times. No human, high or low, rich or poor, seems to possess any great degree of certainty any more about any effort put forth in this "external world."*

*I have discovered much, during these periods of receptivity, concerning the reaction of fear, worry, anxiety, and kindred thoughts upon one's emotional system— how easy it is to become emotionally unbalanced without realizing it, unless constantly on the alert, and with the ability to checkmate one's self.*

I believe that a method must be found for enabling humans to recover their balance, inwardly and outwardly, else civilization, as we know it, is destined to suffer a catastrophic decline.

In this mad world of today we are desperately running about, treating effects, with little or no ability to recognize or deal with basic causes.

And since wrong concepts of the mind are constantly being translated into wrong practices in human life—a revealing study of the powers of mind would seem to afford the greatest promise of liberation from any of our present perplexities.

An exact transcription of impressions received tonight is as follows:

| SHERMAN | WILKINS |
|---|---|
| *Strong wind impression . . .* | We did face strong wind on return flight—early morning of 15th. |
| *Something galling you—May be fact that weather over Arctic keeping you from setting forth on the extensive flight planned at time of month when you are coming into most favorable conditions of sun and moon—unsettled feeling—obstruction of some sort—"Forced down" sensation of what seems to be a rather imminent nature . . .* | Correct. Obstruction was clouds—and, of course, Soviet Government's decision not to continue search. |
| *"Point Barrow"—seems you communicating that base today—Morgan on radio, "Weather not so good there."* | We communicating Barrow every day—but this day more than usual. |
| *May be wrong—but can't feel flight today—get a "stopped sensation"—standing by impression . . .* | Learned of decision to stop all search flights from Alaskan side. |
| *Seems like you had dinner with three people tonight—talked with group of seven later.* | Had dinner out, with three visitors at house. Six people all told. |

## Confirmation through *New York Times* News Dispatch from Wilkins of Recent Flight Impressions

March 17, 1938.

For the past week, at different sittings, I have repeatedly received the impression that Wilkins was to embark on his greatest and longest search flight yet—a flight covering between 2200 and 2600 miles.

Today, a news dispatch from Wilkins tells of his having completed, on Monday, March 14th, into morning of Tuesday, March 15, his longest search flight—2650 miles!

On March 7th, I had received the impression: *"This next flight to be your longest to date—around 1300 miles out—and same distance back—need for good visibility on such a flight imperative—and accountable for flight today—until you can be as humanly certain about this as possible."* (However, Wilkins' next flight, following March 7th, was on March 10th, covering 2,080 miles. And it was not until March 14th that he made the flight of 2,650 miles, closely approximating my impression of 1300 miles out and 1300 miles back).

It will be noted that in the same evening's sitting, many times, I will get seemingly contradictory impressions. Momentarily, on occasion, I will not "seem to feel" Wilkins in the air—and then a flash will come through wherein I "see" him in flight. Because of the difficulty of properly gauging the "time element" in this pioneering telepathic experimentation, I have—from the first—been faithfully recording my impressions as they came into consciousness—letting their specific nature, if later proven accurate, speak for the occurrence of genuine extrasensory phenomena.

In the case of this great flight recently completed by Wilkins on Monday night, my very first recording as I sat down that night, was: *"There has been plane activity today . . ."* then I seem to lose this impression in point of time, when Wilkins was even then still in the air, and went on to record: *"but tomorrow into Wednesday, weather permitting, seems to me to mark one of your greatest attempted flights to date—making use of moonlight as well as daylight . . ."*

Later, this same evening, I again had flashes of Wilkins in flight—each impression, it must be remembered, coming to me independently—usually with no relation to one that has come before. And these other impressions, many of them, seem to have been correct pick-ups, judging from the news dispatch, of incidents actually happening on Wilkins' flight.

If "contact" could once be established and held, between mind of sender and receiver, then a continuity of pick-up would be just as possible as in radio communication. With "contact" coming and going, in its present stage of development, some allowances must be made for occasional loss of "sensing" of exact time relation—if the impression itself is accurate.

I have referred repeatedly, in recent sittings, to "hazardous nature" of flights now being undertaken, and on March 15th, the day after Wilkins' record flight, as well as on March 14th, the actual day of flight, I used the very word "hazardous" in connection with my impressions of his activities, only to have Wilkins say in his news report, published March 17th: "The last two hours were the *most hazardous* of our experience. Clouds blanketed Aklavik, and all but the highest mountain peaks in the neighborhood."

And on the Monday night, when Wilkins was then in the air returning to Aklavik, I apparently received an impression of this very dilemma that he mentions, when I recorded:

See another flight turned back because of weather conditions turning against you—and strenuous effort made to reach base before snow and sleet close you in. . . .

"Snow and sleet" certainly implies the existence of clouds—and this was the form that the impression took in recording itself upon my consciousness . . . but here again there must have been a strong thought in Wilkins' mind—that very night—about the time I was sitting—when he and his crew were actually trying to get back to their *base*, Aklavik, just as I had "seen" above.

Many more points of confirmation concerning the flight just made will be apparent in a study of the news dispatch in relation to the impressions received.

One outstanding pick-up, recorded by me Monday night, the day of his flight, and while he was still in the air, was as follows: *"See plane circling low over certain area—icy waste with open stretches of water . . . has looked from air as though pieces of wreckage caught in ice near water's edge . . . but feel just unusual formation . . . impression 86°—115—location . . ."*

It is exceedingly interesting to note that, while I did not get the impression of the location of this "open lead" correct . . . I felt impelled to try to record the location . . . which Wilkins had actually made a note of—and so reported in his news dispatch!

*Never, throughout all my recorded impressions,* have I ever described a scene, and felt the urge to record any longitude or latitude in connection therewith—except in the case of the lost Russian flyers, a recording which is obviously not in the same category.

This should clearly demonstrate that I was "seeing" the same open lead, or a section thereof, that Wilkins was seeing and charting as he flew over it, and which he describes in his news account thus:

"Since our flight on March 10, many of the leads have closed, and there has been considerable pressure on the ice; but the *most interesting* of our observations was an *open lead* varying from 20 to 500 yards wide and running for about 150 miles along our course between *Lat. 81 and 84 N. and Long. 122 to 127 W."*

My impression of location, 86°—115—was a flashing sensation, at least, of figures Wilkins had put down in relation to this "open stretch of water." I had "seen" this in a pick-up from his own mind, while the flight was still being made.

I think any reasonably minded observer would grant this conclusion upon a study of the facts involved, the exact synchronization of the "time element" with the receiving of this specific impression, including my impulse to set down the location of the "scene" in terms of latitude and longitude—something that I know next to nothing about . . . nor have I any conscious way of knowing what territory Wilkins is traversing on any flight or flights in the far North.

A summary of my impressions checked against Wilkins' news dispatch is as follows:

### SHERMAN
*(March 7th)*

*This next flight to be your longest to date—around 1300 miles out—and same distance back—need for good visibility on such a flight imperative and accountable for flight delay . . .*

*(Previous impression almost exact, except 2600 mile flight was made on second flight following, instead of "next.")*

*(March 10th)*
*You seem to be about to make one of the greatest search flights of its kind ever attempted in the north—a flight which will require the combined light of the sun and moon to supply visibility, and will give you the opportunity of surveying, observationally, a wide and carefully charted expanse of Arctic ice—services of both pilots will be required on this flight, changing off—I get a feeling you will probably be in air between 15 and 18 hours—covering between 2200 and 2600 miles . . .*

*(March 14th)*
*. . . hazardous position—see another flight turned back because of weather conditions turning against you—and strenuous effort made to reach base before snow and sleet close you in . . .*

### WILKINS
### NEWS ACCOUNT
(Published, *N. Y. Times,*
March 17, 1938)
2,650 MILES COVERED BY
WILKINS IN FLIGHT
Visibility is Good Throughout
Most of 19 1/2 Hour Search for
Soviet Airmen

By Sir Hubert Wilkins AKLAVIK, N.W.T. March 15—Yesterday we completed the longest flight we have yet made in our search for Sigismund Levanevsky and his Soviet companion fliers, who were lost last August during a flight from Moscow to Fairbanks, Alaska.

Starting from Aklavik at 7:30 A.M. Pacific Standard time, we flew about north by east, 1,325 miles to reach Lat. 87 degrees N., Long. 90 degrees, W., a point within 200 miles of the Pole. With the exception of a few miles near Lat. 90 on the outward course, the visibility was the best we had yet. Even at Lat. 80 degrees N. we could see the ice, although conditions were somewhat hazy.

*We were in the air 19 1/2 hours.* (Mileage as previously stated—2,650 miles.)

The last two hours were also the most hazardous of our experience. Clouds blanketed Aklavik, and all but the highest mountain peaks in the neighborhood.

Both our radio aids to navigation, the radio direction finder apparatus in the

## SHERMAN

*Work on plane tonight—radio activity—another contact with Iversen . . .*

*See plane circling low over certain area—icy waste with several open stretches of water—has looked from air as though pieces of wreckage caught in ice near water's edge—but feel just unusual formation—impression 86°—115—location . . .*

*There has been plane activity today (and here my sense of time slightly confused, as I went on to record)—but tomorrow into Wednesday, weather permitting, seems to me to mark one of your greatest attempted flights to date—making use of moonlight as well as daylight—feel some such plan strongly in mind . . .*

*(March 15th)*
*You are entering what appears to be a hazardous period for you—feel it necessary to urge caution . . .*

## WILKINS

plane and our radio compass, were ineffective. However, our *short-wave radio* was functioning, and our radio operator was able to keep us informed of the conditions on the ground.

Since our flight on March 10, many of the leads have closed, and there has been considerable pressure on the ice; but the *most interesting* of our observations was an *open lead* varying from 20 to 500 yards wide and running about 150 miles along our course between Lat. 81 and 84 N. and Long. 122 to 127 W.

[There *had been* plane activity that day, March 14th—Wilkins was still in the air as I was sitting—with the above impressions coming to me, exactly relating to this flight. And he was in process of making the *greatest flight* to date, as "foreseen" March 10th and again sensed on night of actual flight.]

[This "plan of flight" could not have been stronger in mind than during time of its actually being made.]

[Repetition of hazardous impression—one hazard already having been encountered on this flight as recorded by me that night.]

## Test 64

March 17, 1938.    11:30–12:00 P.M.

I came home tonight from the lecture at the Psychic Research Society just in time to keep my telepathic appointment with Wilkins.
I have been under really terrific pressure for the past ten days again—physically

and mentally—with an impingement of the economic. All of these factors—intensive creative activity, stomach and intestinal upset, and personal situation dealing with money matters on certain transactions—have had their effect on my consciousness.

I feel in all fairness and honesty that these personal equations should be stated—to be taken into consideration when judging accuracy of impressions received under apparently most favorable of mental and physical conditions—as against times when I am being bombarded with circumstances within and without myself of a disturbing nature.

I should like to know myself, when these tests are completed, whether—despite what might be happening in my own personal life—I have been able, with consistency, to clear my consciousness and continue to "pull through" a good average of verifiable impressions.

Anyone who has attempted to use these "higher powers of the mind" will realize the difficulty encountered in relaxing, and making the conscious mind passive while under pressure. This is particularly true when acute pain is being suffered . . . or when intense mental concern or attention has had to be given to something. Despite even these conditions, with disciplining, a certain mental control can be attained, and held for a time.

An exact transcription of tonight's impressions is as follows:

## Confirmation by Wilkins of Impressions Received

*(This last batch of recorded impressions was picked up by Sir Hubert Wilkins upon his return to New York, and carried with him to Los Angeles, where he had gone by plane to visit Lincoln Ellsworth. This material was airmailed back, arriving at the City Club, Sunday, April 3rd. I received impression I had mail from Wilkins awaiting me at Club and phoned to confirm impression, being told by Joe, the night doorman, that a large envelope was in my box, postmarked "Los Angeles." I made a special trip downtown to get it.)*

| SHERMAN | WILKINS |
|---|---|
| *For a reason that I can't determine at the moment, I get an exhilarating feeling from you as I sit down tonight—an impression—as though you feel particularly good about something tonight—somewhat excited or lively atmosphere—others of crew seen to reflect some of same reaction . . .* | All members of expedition, including myself, *exhilarated* tonight after full day's work preparing to leave. We were to leave Aklavik next morning for Edmonton.<br><br>*Correct.* |
| *Feel you are on the last leg flight concentration now—with several unexpected* | |

| SHERMAN | WILKINS |
|---|---|
| *and somewhat* spectacular *things due to happen—some of which have been pointed by past impressions—Little confusion in mind between flights actually made and flights planned since your mind so strongly on both . . .* | Was thinking strongly of flights since I very much wanted to make two more—*the last one right across the Arctic to Russia.* |
| *More flight action into Friday— Cheeseman—seem to see at controls— grim expression—clouds—Throughout past week of sittings, must finally confess strange mental reaction—conscious interference against receiving of another accident impression—result of bad weather closing in, making safe return from Arctic wastes very hazardous . . .* | Friday, we flew from Aklavik to Edmonton, non-stop, 1,600 miles against a high head wind (9 1/4 hours).<br><br>For the first four hours over clouds, *then one hour in clouds*—then fairly clear. |
| *"Flying blind" impression—fleeting, as though pertaining to future flight . . .* | *Flying blind on 18th for one hour.* |
| *Busy week-end ahead—much action next few days, weather permitting, which I feel it will, for most part.* | Busy all week-end, changing from skis to wheels. |

## Test 65

### March 21, 1938.    11:30–12:00 P.M.

It has been some time again since I have had any direct confirmation from Wilkins on more recent impressions received. Again this has been a period when it has been impossible get mail in or out; he has been so busy in the North, and radio communication has been so limited—with ordinary "traffic" about expedition matters permitting little or no checking through Iversen on tests of a more specific nature.

In one way I am glad of this, since past results have proved that I have done my best telepathic work when my conscious mind has been freed of any supposed information concerning Wilkins' activities—such information, for instance, as might be contained in a news account, however confirmatory that news account may have proven with regard to past events, telepathlcally received by me.

Herewith is an exact transcription of tonight's impressions; received against the *conflict of a severe physical disturbance affecting my stomach and solar plexus—a condition which has been plaguing me, off and on, for several weeks, and to which I felt impelled to refer in a recording of these impressions:*

| SHERMAN | WILKINS |
|---|---|
| *Feeling now your thoughts are on finishing up and getting away—two more flights seem in offing . . .* | Thoughts were often on finishing up—and frequently thought strongly about the *two* flights I wished to make. |
| *Communication Russian Government regarding future plans and contemplated completion flights when search for Levanevsky and comrades is to be abandoned, if unsuccessful.* | First communication about conclusion of flights came on 15th. On 17th came official confirmation. |

# Test 66

## March 22, 1938.  11:30–12:30 P.M.

Today I had the sudden impression that some more mail had arrived for me at the City Club from Sir Hubert Wilkins, and phoned the club and found this was true. I have been busy at home, writing a fifteen thousand word baseball novelette since last Friday.

I made a special trip down to the Club to get the two envelopes of material Wilkins had annotated and returned, finding many interesting confirmations, as usual.

I was home all evening, and an exact transcription of my impressions is as follows:

| SHERMAN | WILKINS |
|---|---|
| *"Packing up" thought comes to me—equipment you no longer need—arrangements being made for supply plane to aid in carrying some of equipment and supplies back when expedition job completed—pick-ups to be made from Point Barrow and Aklavik . . .* | We were packing up at Edmonton at that time. Soviet Embassy offered provide a supply plane, but I replied that it was not required at Aklavik. They did send a plane to Barrow about this date to pick up supplies from there. |
| *Your base stations and crew or men you are dealing with there are being notified* | Men received notice on March 1st that their period of service would terminate |

| SHERMAN | WILKINS |
|---|---|
| *of approximate time you planning to finish up search efforts and pull out of north region . . .* | on March 31st. I intended leaving, or having the men leave on March 21st.<br><br>[NOTE: This impression was received night of the 22nd of March, day after date Wilkins intended to conclude work.] |
| *Seems to me one particular flight you still want to make—feel you turning over in your mind as though you have hunch of some sort you would like to investigate before giving up search entirely . . .* | Had strongly in mind the flight along Canadian islands. |
| *Wheels impression—contemplation of changing back from skis to wheels in week or so—preparatory to flying plane back—turning over to Russian government.* | We were adjusting wheels that day at Edmonton—Wheels were fitted on Sunday, 20th, and the machine flown from the ice to airport. |

## Test 67

### Extra Sitting
### March 23, 1938.    11:30–12:00 P.M.

I had every intention of attempting to finish the baseball novelette upon which I have been working when I got home from downtown tonight . . . but persistent thoughts of Wilkins kept crowding in to disturb my creative thinking.

At 11:30 P.M., though tonight, Wednesday, is an "off night" for telepathic test purposes, I said to my wife that I had an almost uncontrollable urge to sit for impressions.

My wife encouraged me to do so, since this was the first time, in all the months that the tests have been carried on, that this impulse has seized me. And, because the "urge" did seem unusual, I asked her to sit in my study with me, and to be a witness to my recordings. This she did.

I can only explain this difference in my feelings by saying that it seemed to me I had received and was carrying in my subconscious mind, impressions of happenings that were pressing upon my conscious mind, trying to get through to me. I could not get "peace of mind" until I had sat down, and made myself receptive, and had gotten these impressions "out of my system."

Wilkins' confirmation of many of these impressions provides a very interesting *check-back on the past few sittings, wherein I had received a knowledge of some of the changes taking place.*

*Apparently a change in Wilkins' mental attitude made for better reception at my end. It will be noted that I was impelled to stress my own great mental relief . . . also physical . . . as these impressions came through.*

*The following is an exact transcription of impressions received:*

| SHERMAN | WILKINS |
|---|---|
| *Strong, indescribable urge to sit tonight despite fact it is not regular appointed night of week—Have been plagued all day, off and on with strong thoughts of Wilkins coupled with inner feeling of growing mental and physical relief—My thoughts go back to impression received week or so ago of Wilkins feeling exhilarated about something, this feeling reflected by entire crew— and have recorded since my feeling that Wilkins had been trying to impress me with some particular thoughts concerning his activities . . .* | Again all crew exhilarated for we were to leave Edmonton next day.<br><br>We all at cocktail party with many friends. |
| *Now it comes to me—this feeling of exhilaration—the impression I got some days ago—pertaining to your leaving Aklavik—breaking up camp—That's why rest of crew were exhilarated, too— would not surprise me to learn that you are all in Edmonton—that you've been there for several days—perhaps since last Friday—I feel relieved as I record this— as though at last I've pulled through what you've repeatedly been trying to send me telepathically . . .* | |
| *I get feeling of your being back in hotel—Windy trip . . .* | Last Friday we had to fight head winds of from 27 to 47 miles per hour, all way from Aklavik to Edmonton. |
| *Wheels on plane—believe they have been put on now . . .* | [NOTE: This fact confirmed by Wilkins in previous notation, when impression concerning "wheels" first received, night of March 22nd.] |

| SHERMAN | WILKINS |
|---|---|
| *Have you flown on today to Winnipeg? Seem to feel flight action—as though you heading on toward New York...* | We preparing to fly Winnipeg, Thursday—New York, Friday. |
| *Feel you've covered over twenty thousand miles searching for lost Russian fliers...* | Approximately twenty thousand miles during second search. |
| *Don't get final feeling from you about fate Russian fliers—feel you not quite satisfied—that there's a search flight you would like to have made...* | Correct. |
| *Break up ice, Aklavik—spring impression—I got this few nights ago—seems to be partial reason why search ordered abandoned by Russian government—too late in season—warmer weather...* | Break up of ice at and near Edmonton fully responsible for abandoning search. |
| *Relief mentally and physically is enormous—I'm feeling better as I write—Feel much less tension from you—which has been terrific last few weeks...* | Correct—for me. |
| *Yes, I can't explain how I feel it—but I'm sure I'm right about your not being in Aklavik. I seem to see your base there deserted—and some man, Point Barrow, getting ready to leave there by plane also—You've been to some dinner occasion tonight—wined and dined and made a speech, recounting your experiences...* | Correct.<br><br><br><br>Correct—Lunch and dinner. |
| *You've thought of me in more relaxed way past few days and with different emotional urge—you've gone over additional recorded impressions of mine picked up at Edmonton—see you sitting in hotel room, checking them—You'll either bring them on to New York with you, or put some in mail...* | Correct—on Wednesday and Tuesday.<br><br><br>Mailed. |
| *Would not surprise me to see you in New* | *Proved correct.* |

| SHERMAN | WILKINS |
|---|---|
| York as early as Saturday—I get feeling your eagerness to be about other things—see Ellsworth—check matters regarding new submarine . . . | Correct. |
| Equipment being shipped—impression—supply plane to pick up some of it . . . | Equipment being shipped. |
| Believe you are coming on to New York in search plane. | Correct. |

# Test 68

## March 24, 1938.     11:30–1200 P.M.

Tonight, the Strath-Gordons returned with us from the Open Forum meeting of the Psychic Research Society, and Dr. Strath-Gordon sat with me during my half-hour period.

I found my feelings unchanged—that Wilkins had finished his research work, and was en route back to New York. My own freedom from mental and physical tension was again marked, which I believe has a direct association with a changed condition surrounding Wilkins.

An exact transcription of impressions received is as follows:

| SHERMAN | WILKINS |
|---|---|
| Feeling just as strong tonight that you have left Aklavik some days ago—first to Edmonton where skis changed to wheels—then—yes, I feel you are now in Winnipeg—Fort Garry Hotel—I now know name of this hotel—believe you stopped MacDonald in Edmonton, name of hotel I also since have come to know . . . | Stopped at Fort Garry on *night of 24th*. Left there on 25th. Correct. |
| I hear you making address in Edmonton—seems that you thanking different countries for cooperation which made search flights possible—Canada, United States of America, and Soviet Russia . . . | Correct. Spoke of value of our activities in relation to international friendship and cooperation. |

| SHERMAN | WILKINS |
|---|---|
| *I seem see mental picture of Russian fliers walking toward shore across ice—Did you say in your talk you consider it possible they still could be alive, and making away across drifting ice floes toward land? Drift would be tremendous by this time—Feel this is speculation in your mind—though strong probability Russian flyers have perished . . .* | Yes.<br><br><br><br><br>Correct. |
| *Think these comments made at luncheon club at Edmonton, and repeated in address at Winnipeg.* | Correct. We arrived at Winnipeg just in time to attend a *luncheon* of the Aviation Section of Department of Commerce. They did not know we were coming until we actually arrived at airport where luncheon was to be held. Ten minutes after landing, I was giving a talk. |
| *The nerve tension which I have had to combat for past month or so is almost completely gone, and I seem to see these mental pictures clearly, as though obstruction removed—"Feeling" contact with your own consciousness seems of different, clearer nature—will be interested to discover if this is actually true . . .* | Correct. Mental strain of not being able to fly more often was over. |
| *I see welcoming committee at landing field Winnipeg to greet you—feel Kenyon at controls flight from Aklavik to Edmonton—also en route to Winnipeg . . .* | Correct. |
| *See Cheeseman and Dyne, I believe, seated at soda counter, tanking up on ice cream sodas . . .* | Not while the Scotch lasted. |

(This sitting proved to be the last session of the five months' series.)

## Letter to Dr. Gardner Murphy
## Department of Psychology Columbia University

April 4, 1938
Dr. Gardner Murphy,
Department of Psychology,
Columbia University,
New York City.

Dear Dr. Murphy:

I am herewith enclosing the last of the annotated impressions returned to me by Wilkins following completion of his six months' Arctic search for the lost Russian flyers.

I am sorry that we were unable to perform more of the ESP card tests desired by you, but conditions in the North with Wilkins made this impossible.

Sir Hubert was compelled to fly to the coast immediately for a conference with Lincoln Ellsworth. He will be back around April 11th, at which time I should like to arrange a meeting of the few principals associated observationally with these extrasensory perceptive experiments.

I feel that I should report to you the great mental and physical relief experienced by me at the termination of these telepathic tests. Keeping up this program, three nights a week, for this length of time, with all other activities, proved more of a strain than I could anticipate. A check on all these factors should some day prove most interesting.

Let me thank you profoundly for your own fine interest, and the permission granted me to mail you copies of my recorded impressions the morning following their reception, long before I could secure any confirmation.

I have been thoroughly appreciative of your own over-burdened activities during this period, but you have been most helpful in enabling me to have all this work on record with you . . . and when you are finally granted time to make an evaluation of these tests, I shall wait your report with interest.

Sincerely—

*Harold M. Sherman*

# Hampton Roads Publishing Company

*... for the evolving human spirit*

Hampton Roads Publishing Company
publishes books on a variety of subjects,
including metaphysics, health,
visionary fiction, and other related topics.

For a copy of our latest catalog, call toll-free
(800) 766-8009, or send your name and address to:

Hampton Roads Publishing Company, Inc.
1125 Stoney Ridge Road
Charlottesville, VA 22902

e-mail: hrpc@hrpub.com
www.hrpub.com